Fungal Protoplast

A Biotechnological Tool

Fungal Protoplast

A Biotechnological Tool

D. Lalithakumari

Science Publishers, Inc.

Enfield (NH), USA Plymouth, UK

SCIENCE PUBLISHERS, INC.
P.O. Box 699
Enfield, New Hampshire 03748
United States of America

Internet site: *http://www.scipub.net*

sales@scipub.net (marketing department)
editor@scipub.net (editorial department)
info@scipub.net (for all other enquiries)

Library of Congress Cataloging-in-Publication Data

Lalithakumari, D.
Fungal protoplast : a biotechnological tool/D. Lalithakumari.
p. cm.
Includes bibliographical references and index.
ISBN 1-57808-093-2
1. Fungi—Biotechnology. 2. Fungal protoplasts. I. Title.

TP248.27.F86 L35 2000
660.6—dc21

00-037192

Published by Science Publishers, Inc., USA.
Printed in India

Foreword

The first experiments involving protoplasts date back to almost fifty years. Nevertheless its age, protoplasts remain modern and indispensable tools in current genetics. Whereas in the early days protoplasts were used solely for biochemical and biophysical investigations, in recent times protoplasts have become very important instruments in classical and molecular genetics. This change occurred about 25 years ago, when we showed that protoplasts can be fused. These experiments were designed to create new variants and species able to produce novel interesting metabolities. Since that first experiment, protoplast fusion has become a major approach in strain improvement programmes of industrially important fungal species such as antibiotic producers, biocontrol agents and edible mushrooms. The technique of protoplast fusion is a means to combinatorial biosynthesis, now routinely in use for the biosynthesis of novel metabolities, not produced by the parent strains. A new impetus of the use of protoplasts came, when it was shown that protoplasts can be easily transformed and transfected by DNA, making them to indispensable in molecular genetics.

Writing the present book *Fungal Protoplasts* was a brave and interesting initiative of Prof. Dr D. Lalithakumari. This book collects all the fungal protoplast work published so far, starting from the early days up to the most recent data. It contains a detailed collection of the results obtained during the long history of the fungal protoplast work that has been published for different species in the literature, from conditions and enzymes to prepare protoplasts, media for regeneration and conditions for fusion, from microscopic and physiological description to genetic analysis and novel biochemical properties of the obtained recombinants and hybrids. This overview is supplemented with a massive research work for the improvement of biocontrol agents carried out in her laboratory. This work provides not only an overview of the literature as such, but also practical conditions to carry out the work.

Therefore, this book is without doubt of interest for all those who are interested in fungal protoplast fusion as a means to strain improvement.

Prof. Dr Jozef Anne
Rega Institute
Katholieke Universiteit Leuven
Belgium

Preface

This book has been written primarily as a guide for young scientists and students of molecular biology and biotechnology who have novel ideas for the improvement of animal, plant and microbial strains. A rigourous effort has been made to achieve this objective through thorough coverage and sufficient discussion to enable easy understanding of the subject. I have tried to include up-to-date literature on the active utilization of protoplasts in modern biology. This book, with extensive coverage of important information under one cover, therefore will be a ready reckoner for students and researchers in life sciences. The flood of new ideas and experiments in the modern recombinant DNA techniques and applications using protoplast highlight the importance of protoplasts as a biotechnological tool. Although protoplast fusion technology has immense possibilities in pure and applied genetics its application has been limited on account of limited information available on living protoplasts. The scattered literature on the subject has been carefully screened and included in this book for background information. This is the only book of this nature on fungal protoplasts.

The subject matter has been presented under different chapters in this book. The isolation of purified and physiologically viable protoplasts with precise factors governing the release of protoplasts are described in Chapter 1. Chapter 2 deals with the importance and the art of regeneration and reversion of protoplasts to their parent culture. The conditions for intraspecific, interspecific and intergeneric protoplast fusion and subsequent selection of fusants (heterokaryons, diploid and recombinant strains) are described in Chapter 3. Chapter 4 explains the challenges of protoplast fusion in biotechnology for strain improvement, production of recombinants, super strains etc. A broad outline is given at the beginning of each chapter offering the readers an overview of the subject. I sincerely hope that I have succeeded in transmitting the excitement generated by our increasing awareness of the potential of fungal protoplasts as biotechnological tools.

D. Lalithakumari

Acknowledgements

I am deeply indebted to Prof. J. Anne, Rega Institute, Catholic University, Leuven, Belgium who instantly responded to my request and made available all his review articles and several of his publications. Anne and Peberdy's contributions in the area of protoplasts research set the stage for other workers.

I gratefully acknowledge the support received from Drs de Vries and Wessels, The Netherlands, who readily responded and sent their publications on protoplast research, especially with *Schizophyllum commune*. Dr G. J. Boland, Department of Environmental Biology, University of Guelph is duly acknowledged for his valuable help in providing his papers on protoplast regeneration in *Sclerotinia sclerotiorum*.

My past and present students have helped me in this endeavour in many ways and every brick for construction of this book is their valuable contribution. My first work on the isolation of protoplast of *Pyricularia oryzae* (*Magnoporthe grisea*) with Dr Saradha Kumari was followed by Dr P. Annamalai to whom I am deeply indebted. My thanks are due to Dr R. Revathi and Dr P. Vijayapalani who have contributed to most of the research work on protoplasts of *Venturia inaequalis*. Mrs S. Karpagam and Miss R. Kalpana have contributed to the morphological and chemical variations in the regenerated protoplast cultures *Trichoderma* sp. and *Colletotrichum capsici*. Mrs C. Mrinalini and Mr G. Chellappa have done outstanding research on the strain improvement of *Trichoderma* spp., the former, proving the potentiality of a superior antagonistic strain and the latter about a superior strain producing high cellulase and chitinase enzymes. Dr A. Elavarasan has developed a method to mutate protoplast in *Rhizoctonia solani*. His protoplast mutants produced fruiting bodies under chemical (iprodione) stress. My thanks to all of them not only for their research contributions but also for their whole hearted co-operation along with Mr R. Sridharan, Mr S. Chandrasekaran, Dr V. Kaviarasan, Mrs K. Subhashree, Miss S. P. Kamala Nalini and

Mrs V. P. Sobha Kumari in the preparation of reams of manuscripts, patient proof reading of immeasurable 'miles' of writing.

I profusely thank Prof. A. Mahadevan, former Director, CAS in Botany, University of Madras for his academic and scholastic support given to me in writing this book.

The financial support received from the University Grants Commission for writing this book is gratefully acknowledged.

Contents

CHAPTER 1

Isolation of Fungal Protoplasts of Filamentous Fungi

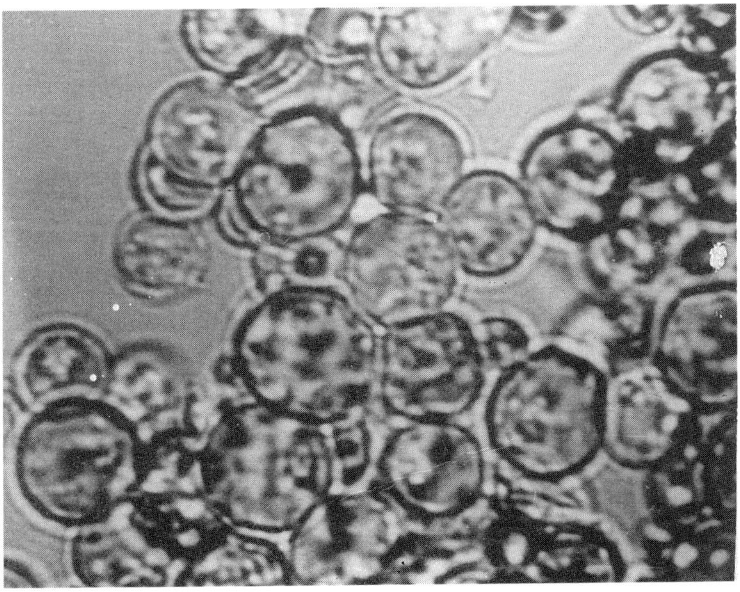

The term protoplasts (Weibull, 1953) has been used for both microorganisms and higher plant cells to indicate the structure remaining after complete removal of the cell wall. Unlike the plant cells, the absence of cell wall remnants at the plasmalemma in filamentous fungi, the term protoplast is used to describe cells obtained after digestion of the cell wall and fulfilling the criteria as described by Villaneuva and Garcia Acha (1971) i.e. 1. release of protoplasts through one or several pores, leaving empty cell wall behind, 2 loss of rigidity, resulting in a spherical form and 3. osmotic fragility.

1.1 ISOLATION OF FUNGAL PROTOPLASTS

The protoplasts limited by the cell membrane can be isolated by mechanical or enzymatic removal of the cell wall. The isolated protoplast is only a naked cell surrounded by plasma membrane. A viable protoplast is capable of cell wall regeneration, cell division, growth and also revert to the parent culture.

The cell wall not only offers rigidity but also protection against external factors. It has to be removed without damaging the cell membrane to obtain intact protoplasts. The protoplast also applies an equal but opposite pressure upon the cell wall as a result of which there is a balance of pressure. When the cell wall is removed, the naked cell not only loses its shape but also is susceptible to external environment leading to plasmolysis and disintegration of protoplasts. Suitable osmotic stabilizers protect the naked cells from plasmolysis.

The cell wall can be removed either by digestion using different lytic enzymes or through mechanical rupture. The lytic enzyme is not common to all filamentous fungi and the combination varies from one organism to another depending on the chemical composition of the cell wall. Fungal walls, like those of plants and bacteria, consist of a rigid layer outside the protoplast whose major constituent is chitin, which protects from osmotic and other changes in the environment. Many dimorphic fungi are known to have altered wall composition at different stages of development, which are responsible for the characteristic shape of the cell and have to be modified when the cell changes during the growth of the hyphal tip. The hypha extends and branches out to produce conidiophore bearing conidia and sporangiophore producing sporangia. Besides, various reproductive structures like ascospores, basidiospores, sclerotia, chlamydospores etc. are formed according to the class of fungi and stages of development.

There are two different methods for the isolation of purified protoplasts. They are 1. non-enzymatic or mechanical rupture of cell wall and 2. enzymatic lysis of the cell wall.

a. Non-Enzymatic Procedure for the Isolation of the Protoplasts

This procedure is economical when compared to the cost of commercially available enzymes or the time involved in the laboratory production of enzymes. This procedure dispenses with the need for a lytic enzyme. The mechanical rupture of the cell wall is carried out using a French pressure cell (FA 079). The pressure exerted (lb/inch^2) at 4°C and the exposure rate vary with different fungi.

Use of 2-deoxy-D-glucose and magnesium sulphate

This method involves the culture of cells in the presence of 2-deoxy-D-glucose and high concentrations of magnesium sulphate. Adoption of this procedure may need caution as 2-deoxy-D-glucose causes a reduction in the respiratory activity and the effect was also distinct on other metabolic activities of the regenerated protoplast. This method cannot be used for most filamentous fungi.

Use of insect tissue culture medium (Tyrell & MacLeod 1989)

This is a highly specialized process of protoplast release adopted for fungi belonging to *Entomophthora*. Conidia of entomogenous fungi inoculated into insect tissue culture medium produced germ tubes that later released their cytoplasmic contents as protoplast. The protoplast thus isolated showed several differences from the normal protoplast. They were spindle shaped, but osmotically fragile, attaining a spherical form and bursting when the medium was diluted. Also these protoplasts are capable of proliferation and following a period of incubation, they aggregate to form hollow spherical structures. Though this pattern of development is a regular one and has been interpreted as a part of the life cycle of the fungus in its insect host, this method cannot be adopted successfully for other fungi.

b. Enzymatic Procedure for the Isolation of the Protoplasts

The use of protoplasts in many areas of biochemical, morphological, physiological and genetical studies demand large-scale production of purified and physiologically viable protoplast. Although mechanical and other non-enzymatic methods have been reported, their use is limited as most of them are specific for a particular organism. Besides, due to possible concern about the physiological alterations that might be induced in the protoplast by these treatments, most workers favoured the use of lytic enzymes to isolate protoplasts. The isolation of protoplasts from fungi using lytic enzymes was initially used in preparing cell free extracts and organelles for biochemical studies recently the use of protoplasts for fusion and DNA transformations has made it an important tool in genetic studies. Availability of large numbers of protoplasts depends on the suitable lytic enzymes. The first enzyme described to dissolve fungal (yeast) cell wall was the digestive juice of the snail (*Helix pomatia*) in 1914, though it was not until 1957 that its lytic properties were used to induce protoplast formation in yeast (Eddy and Williamson, 1957). Since then, protoplasts have been prepared from species representing all major taxonomic groups belonging to different classes of Zygomycotina, Ascomycotina, Basidiomycotina and Deuteromycotina. The snail digestive juice preparation was reported to be effective against *Neurospora crassa* (Emerson and Emerson, 1958; Bachmann and Bonner, 1959) *Cephalosporium acremonium* (Fawcett *et al.,* 1973), *Aspergillus nidulans* (Ferenczy *et al.,* 1974) and *Geotrichum candidum* (Ruiz-Herrera and Bartnicki-Garcia, 1976). An enzyme derived from *Cytophaga* is also effective against *C. acremonium* (Fawcett *et al.*, 1973; Hamlyn and Ball, 1979). The list of organisms showing mycolytic activity is increasing day by day considering the diversity of cell wall composition in filamentous fungi. Subsequently other workers used the same enzyme extensively with different yeasts. This enzyme is named as helicase, sulphatase and glusulase. A second enzyme zymolase derived from *Arthrobacter luteus* is used for protoplast isolation from bakers yeast. Various actinomycetes produce enzymes that have lytic activity against yeasts, *Irpex lacteus* (driselase), *Penicillium funiculosum* (cellulase) *Oxysporus* Sp. (cellulase) and other *Streptomyces* spp. produce effective lytic enzymes. Isolates of *Streptomyces* and *Micromonospora* are most commonly used although fungi such as *Penicillium purpurogenum* (Beggs, 1978), *Trichoderma harzianum,* and *T. Viride* (Benitez *et al.,* 1975) are equally useful.

The composition of the lytic enzyme that is effective in protoplast release has been investigated by many researchers. Sietsma *et al.* (Binding and Weber, 1974) showed that cellulase from *Laminaria* sp. was essential for protoplast release from *Pythium* sp. In *Schizophyllum commune* (Brinboim, 1971) chitinase and S-glucanase (1, 3-glucanase) were required to release protoplasts (De Vries and Wessels, 1972). The list of lytic enzymes markedly increased in recent years. The most important enzymes currently used are cellulase, β-1, 4-glucanase, β-glucuronidase, chitinase, pectinase, helicase, driselase, zymolase and Novozym 234. Mycolytic enzyme preparations generally contain a variety of enzyme activities such as α, β-D-glucanases, chitinases, proteases etc. Chitinase has been applied in the effective release of protoplasts from a number of fungal species, containing a significant amount of chitin in their cell wall.

No single enzyme is effective in the lysis of the cell wall. It is due to the fact that the cell wall components of fungus are resistant to individual and combinations of enzyme mixtures. Period of treatment and concentration are the critical factors and have to be standardized for a particular fungus.

c. *Combinations of Enzyme Mixture*

A combination of enzymes have been reported to release the maximum number of protoplasts. A list of enzymes/combination of enzymes and stabilizers is presented in Table 1.

Table 1 Conditions for protoplast isolation from various fungi

Organism	Enzyme/enzyme combination	Stabilizers	References
Aspergillus nidulans	Novozym 234	0.6 M KCl	Hamlyn *et al.* (1981)
Aspergillus niger	Cellulase CP		
Acremonium chrysogenum	Lytic enzyme L1	0.7 M NaCl	Hamlyn *et al.* (1981)
Penicillium chrysogenum	Novozym 234	0.7 M NaCl	Hamlyn *et al.* (1981)
Aspergillus parasiticus	β-glucuronidase	1.0 M NaCl	Tyagi *et al.* (1981)
Trichoderma matsutake	Cellulase R-10 + Zymolyase 5000 + β-glucuronidase	0.6 M NaCl	Abe *et al.* (1982)
Rhizoctonia solani	Cellulase Onuzuka R-10 + Macerozyme + β-glucuroidase	0.6 M Mannitol	Hashiba & Yamada (1982)
Aspergillus nidulans	Chitinase + β-glucuronidase	0.6 M KCl	Issac & Gokhale (1982)
Gibberella fujikuroi	Chitinase + β-glucuronidase	0.8 M Mannitol	Harris (1982)
Cochilobolus heterostrophus	β-glucuronidase	0.7 M KCl + 0.8 M Mannitol	Leach & Yoder (1982)
Gibberella zeae	Chitinase + β-glucuronidase	9.6 M KCl	Lesile (1982)
Coprinus cinereus	Cellulase onuzuka	0.5 M $MgSO_4$	Akmatsu *et al.* (1983)
Collybia velutipes	Cellulase onozuka + Zymolyase		Yamada *et al.* (1983)
Trichoderma reesei	Driselase	0.5 M KCl	Picataggio *et al.* (1983)
Verticillium albo-atrum	Helicase + Mycolase	15% sucrose	Typass (1983)
Verticillium dahilae	Helicase + Mycolase		Typass (1983)
Trichoderma rubrum	Novozym 234		Srikantha & Rao (1984)
Aspergillus oryzae	Chitinase + β-glucuronidase		Yabuki *et al.* (1984)
Fusarium oxysporum	Novozym 234 + Cellulase CP	0.7 M NaCl	Marriott *et al.* (1984)
Gaumannomyces gramminis	Novozym 234 + Cellulase CP	0.6 M KCl	Stanway & Back (1984)
Curvularia inaequalis	Snail enzyme		Laurila *et al.* (1985)
Trichoderma reesei	Snail enzyme	0.5 M KCl	Laurila *et al.* (1985)
Pleurotus ostreatus	Novozym 234	0.1 M Mannitol	Go *et al.* (1985)

(*Contd.*)

Table 1 Contd.

Organism	Enzyme/enzyme combination	Stabilizers	References
Coprinus macrorhizus	Chitinase + Cellulase + Zymolase	0.7 M $MgSO_4$	Yanagi *et al.* (1985)
Verticillium albo-atrum	Streptozyme	0.7 M $MgSO_4$	Morehart *et al.* (1985)
Fusarium triticum	Novozym 234	1.0 M $MgSO_4$	Lynch *et al.* (1985)
Fusarium oxysporum	Novozym 234	1.0 M KCl	Lynch *et al.* (1985)
Cochiobolus heterostrophus	β-glucuronidase + Driselase + Chitinase	0.7 M NaCl	Turgeon *et al.* (1985)
Geotrichum candidum	Novozym 234	0.8 M $MgSO_4$	Jacobsen *et al.* (1985)
Pyricularia oryzae	Zymolase + β-glucuronidase + Driselase + Cellulase	0.6 M KCl.	Asai *et al* (1986)
Pleurotus ostreatus	Cellulase onozuka + Driselase + β-glucuronidase	0.7 M Mannitol	Toyomasu *et al.* (1986)
Pleurotus cornucopiae	Novozym 234		Lee *et al.* (1986).
Pseudocercosporella herpotrichoides	Rozyme HP 150 + Driselase + Cellulase CP	0.6 M KCl	Hocart *et al.* (1987)
Beauveria bassiana	Cellulase + Chitinase + β-glucuronidase lysozyme	0.6 M $(NH_4)_2SO_4$	Pfeifer & Khachatourians (1987)
Gibberella zeae	Chitinase + β-glucuronidase	0.8 M Sucrose	Adams *et al.* (1987)
Pleurotus sp.	Cellulase RS + Driselase+ Macerozyme R 10 + Zymolase 5000	0.6 M Mannitol	Hrmova & Seltrennikoff (1987)
Ganoderma lucidum	Novozym 234 + β-glucuronidase		Choi *et al.* (1987)
Mycorrhizal fungi	Cellulase onozuka R10 + Driselase	0.7 M $MgSO_4$	Hebraud & Fevre (1987)
Termitomyces clypeatus	Cellulase + Chitinase + Novozym 234	0.5 M KCl	Mukherjee & SenGupta (1988)
Trichoderma harzianum	Novozym 234	0.5 M KCl	Stasz *et al.* (1988)
Schizophyllum commune	Novozym 234	0.5 M $MgSO_4$	Munoz Rivas *et al.* (1988)
Ustilago hordei	Novozym 234	0.5 M $MgSO_4$	Wang *et al.* (1988)
Cercospora nicotianae	β-glucuronidase, $^+$Cellulase	1.0 M Mannitol	Gwinn & Daub (1988)

(*Contd.*)

Table 1 Contd.

Organism	Enzyme/enzyme combination	Stabilizers	References
Neurospora crassa	β-glucuronidase	1.0 M Mannitol	Gwinn & Daub (1988)
Hypoxylon mammatum	Novozym 234	0.8 M Sucrose / Glucose	Griffin *et al.* (1989)
Fusarium solani f.sp. pisi	β-glucuronidase + Driselase + Novozym 234	1.2 M $MgSO_4$	Soliday *et al.* (1989)
Trichoderma reesei	α-1-3 glucanase + β-1-3 glucanase + β-1-4 glucanase + Chitinase + Protease	0.6 M KCl	Sandhu *et al.* (1989)
Aspergillus niger	exo-B 1-4 glucanase + endo-B 1-4 glucanase + β-1-4-glucuronidase		Das *et al.* (1989)
Fusarium oxysporum f.sp. conglutinans	Novozym 234	0.8 M Sucrose	Mamol *et al.* (1989)
Puccinia graminis	Novozym 234 + Cellulase	0.8 M $MgSO_4$	Huang *et al.* (1990)
Botrytis cinerea	β-glucuronidase + Cellulase R 10 + Driselase	0.6 M Mannitol	Braun & Heisler (1990)
Colllectotrichum gloeosporioides	Novozym 234	1.2 M Mannitol	Tebeest and Weidermann (1990)
Phytophthora capsici	Driselase + Cellulase CP + Novozym 234	1 M Mannitol + 7 mM $MgSO_4$	Lucas *et al.* (1990)
Bipolaris virens	β-glucuronidase + Chitinase + Driselase or Novozym 234	1.2 M $MgSO_4$	Ossanna & Mischke (1990)
Dreschslera oryzae	Cellulase + Pectinase + β-glucuronidase + Chitinase	0.6 M Sucrose-Mannitol	Annamalai & Lalithakumari 1991
Pyricularia oryzae	Cellulase + Pectinase + β-glucuronidase + Chitinase	0.6 M Sucrose-Mannitol	Kumari & Lalithakumari (1987)
Cercospora kikuchii	Novozym 234 + β-glucuronidase	0.7 NaCl 10 mM $CaCl_2$ 10 mM $NaPO_4$	Upchurch et al. (1991).
Ustilago maydis	Novozym 234	0.5 M $(NH_4)_2SO_4$	Editmann & Schauz (1992)
Venturia inaequalis	Novozym 234 + Pectinase	0.6 M Sorbitol	Revathi & Lalithakumari (1992)
Venturia inaequalis	Novozym 234 + β-glucuronidase	0.6 M Sucrose 7 0.6 M Sorbitol	Vijayapalani (1995)

(*Contd.*)

Table 1 Contd.

Organism	Enzyme/enzyme combination	Stabilizers	References
Colletotrichum capsici	Novozym 234	0.6 M NaCl	Kalpana, (1995)
Rhizoctonia solani	Novozym 234	0.6 M Mannitol	Elavarasan, (1996)
Trichoderma harzianum	Novozym 234	0.6 M KCl	Mrinalini & Lalithakumari, (1998)
Trichoderma longibrachiatum	Novozym 234	0.6 M NaCl	Mrinalini & Lalithakumari, (1998)
Trichoderma reesei	Novozym 234	0.6 M KC1	Chellappa (1997)
Trichothecium roseum	Novozym 234	06. M KCl	Lalithakumari & Annie, (1998) (unpublished data)

Source: Updated from Annamalai & Lalithakumari, 1993.
Indian Review of Life Science, 13, 41-59.

The release of protoplasts irrespective of the lytic enzymes was similar in all filamentous fungi. In many cases, a combination of enzymes was more effective for maximum release of protoplasts than a single enzyme. The lytic enzyme mixture combination for the release of protoplasts from the sporidia of *U. maydis* was Helicase and Onozuka-R10. This combination effected 50 per cent release of sporidial protoplasts of *U. maydis* in 3 h (De Waard, 1976). In *Penicillium chrysogenum* a complete conversion of mycellium into protoplasts occurred during 5 h incubation with a combination of cellulase and sulfatase or cellulase plus strepzyme 12705 (Anne, 1977).

The mycelial wall of *Coprinus macrorhizus* was effectively digested with a mixture of Chitinase, *(Streptomyces griseus)* Cellulase *(Trichoderma viride)* and Glucanase *(Arthrobacter luteus).*

Twenty commercially available polysaccharase enzyme preparations were screened for their lytic activity against *P. herpotrichoides*. Novozym 234, found to have some lytic effect, was used in paired combinations with other enzymes to screen for increased activity. These enzymes were also tested individually. Enzymes giving increased yields of protoplasts were then tested in novel combinations of two or three enzymes at a concentration of 5 mg/mL of each.

Cellulase (Bhoeringer), Cellulysin, Helicase, Lytic enzyme L1, Macerozyme R-10, Meicelase P, Pectinase (Serva), Pectolyase, Rolament P and snail enzyme (Szeged) were effective for protoplast release either singly or in combination.

The combination of Cellulase, Onozuka R-10 and Driselase was the most efficient mixture for maximum release of protoplasts from the dikaryotic strain *H. cylindrosporum*. Protoplast release in *Eremothecium ashbyii, Trichoderma reesei* and *Penicillium chrosogenum* was achieved using commercially available enzymes, Novozym 234 and Funcelase (Lakshmi and Chandra, 1993). In Bipolaris oryzae (Annamalai and Lalithakumari, 1991) a combination of *Cellulase*, *Pectinase*, *β-glucuronidase* and *Chitinase* was effective and in *Venturia inaequalis* (Revathi and Lalithakumari, 1992) Novozym 234 and Pectinase was effective in releasing the maximum number of protoplasts. High yield of protoplasts of *Sclerotium rolfsii* was obtained using Novozym 234 in 1 ml of 0.05 M maleic acid-NaOH buffer (pH 5.0) containing 0.6 M KCl (Kelkar *et al.,* 1990). *In Trichoderma harzianum and T. longibrachiatum* treatment with Novozym 234 only released maximum number of protoplasts (Mrinalini, 1997). In *Rhizoctonia solani* (Elavarasan, 1996) and *Colletotrichum capsici* (Kalpana, 1995) Novozym 234 was effective in maximum release of protoplasts.

d. Effect of concentration of lytic enzymes

The wall lysis of lytic enzyme activity depends on the concentration of lytic enzymes. De Vries and Wessels (1972) reported the effect of enzyme concentration on the protoplast release in *Trichoderma viride*. The rate of release of protoplasts decreased with the time of incubation in increased concentration of enzymes. The influence of concentration of lytic enzymes directly affects the release of protoplasts. Anne (1977) had shown this with *P. chrysogenum* (Fig. 1).

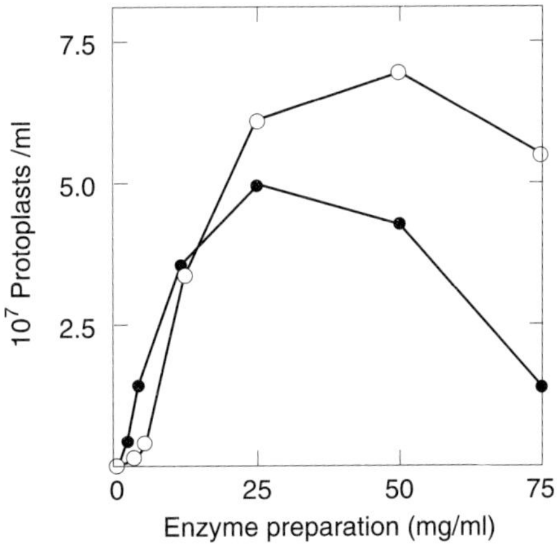

Fig. 1. Effect of concentration of TLE (o) or strepzyme (•) on the yield of protoplasts. Counts were made on 150 mg *P. chrysogenum mycelium* incubated for 3 h in 3 ml of a 0.55 M NaCl solution, containing increasing concentrations of TLE of strepzyme Reproduced from Anne, 1977.

The number of protoplasts increased with increasing concentration of lytic enzymes, but high concentration are sometimes harmful, resulting in the lysis of protoplasts soon after their appearance indicating the toxic levels of the lytic enzyme. Novozym 234 as a lytic enzyme when used at different concentrations (2, 3 and 5 mg/mL) under shaken conditions with *Collectotrichum capsici* (Table 2) yielded maximum protoplasts with 5 mg/ mL Novozym 234 (Kalpana, 1995).

Table 2 Effect of Novozym 234 concentrations on the release of protoplasts of C. *capsici*

Osmotic Stabilizer	No. of protoplasts released × 10^4 Novozym 234 concentration (mg/mL)		
	2	3	5
KCl (0.6 M)	1	3	5
KCl (0.8 M)	1	2	3
NaCl (0.6 M)	4	8	12
NaCl (0.8 M)	2	4	6

Source: Kalpana, 1995. M. Phil Dissertation, Univ. of Madras, 26–34.

For *T. harzianum* and *T. longibrachiatum* 3 mg/mL of Novozym 234 was effective for the release of protoplasts (Table 3).

Table 3 Effect of different concentrations of cell wall lytic enzyme on the release of protoplasts in *T.* harzianum and *T.* longibrachiatum

Strains of Trichoderma	Concentration of enzyme	No. of protoplasts released (× 10^6)
	2 mg/mL	1.1
Th 1	3 mg/mL	2.3
	5 mg/mL	3.1
	2 mg/mL	1.0
T1	3 mg/mL	1.6
	5 mg/mL	2.5

Source: *Mrinalini 1997.* Ph. D. Thesis, Univ. of Madras, 66–102.

e. Effect of Incubation Time with Lytic Enzymes

Incubation time with lytic enzyme is another critical factor for the release of protoplasts. The optimal period of lytic enzyme treatment differs between fungal species and also within the strain i.e. it depends on the age of the culture and the different stages like young germlings, old mycelium, conidia, chlamydospores, sclerotia etc. of the culture. The number of protoplasts released increased to a maximum after 3 to 4 h of incubation in *Penicillium chrysogenum* (Fig. 2) and prolonged incubation did not increase the yield (Anne. 1977).

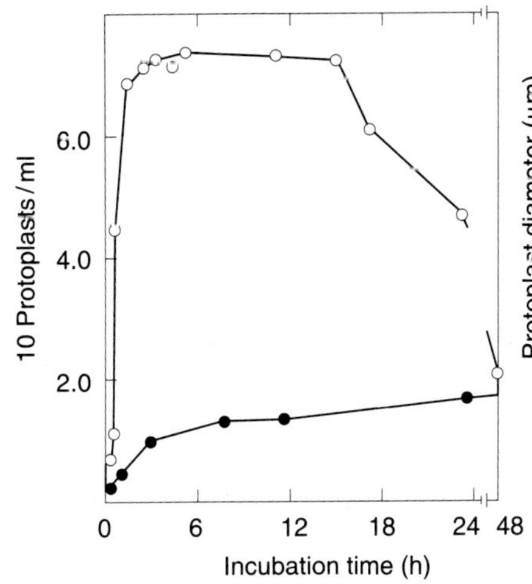

Fig. 2. Effect of incubation time on the number of protoplasts released (o) and on their diameter (•). Investigations were performed on 150 mg (wet weight) of 16 h old *P. chrysogenum mycelium* incubated (27°C, 150 rpm) in 3 l of a solution of 0.55 M NaCl containing 30 mg strepzyme and 60 mg TLE (Reproduced from Anne 1977).

Longer incubation (16 h) resulted in a steady loss of protoplast due to lysis and after 48 h and 96 h only 26.5 and 20 per cent respectively had survived. The size of the protoplasts increased to about 4.8 µ after 1 h to 21.9 µ after 96 h. In the early hours of incubation mostly small sized protoplasts are released. The size of the small protoplasts ranged from 4 to 6 µ and were mostly anucleated. A typical experiment (Fig. 3) when followed sequentially as function of time with *S. rolfsii* gave the maximum yield of protoplasts at the end of 4 h (Kelkar *et al.* 1990).

Kuwabara *et al.*, 1989 have clearly showed the marked effect of the age of the culture on the protoplast release in *Robillarda sp.* Y 20 which is a filamentous fungus producing cellulases (Fig. 4a). Maximum release of protoplasts of *D. oryzae* was obtained (Plate 1, Fig. 4b) after 3 h incubation (Annamalai and Lalithakumari, 1991).

Prolonged treatment after 3 h did not in anyway improve the yield of protoplasts. Rapid liberation of a large number of protoplasts from *E. ashbyii* in 10 min (Fig. 5) appears to be the shortest time reported for the release of protoplasts. In *Curvularia inaequalis* the release of protoplasts was observed in 20 min (Lakshmi and Chandra, 1993).

Protoplasts of *V. Inaequalis* (Plate 1a, Table 4) released 1.5 h after the onset of incubation with Novozym 234 and β-glucuronidase.

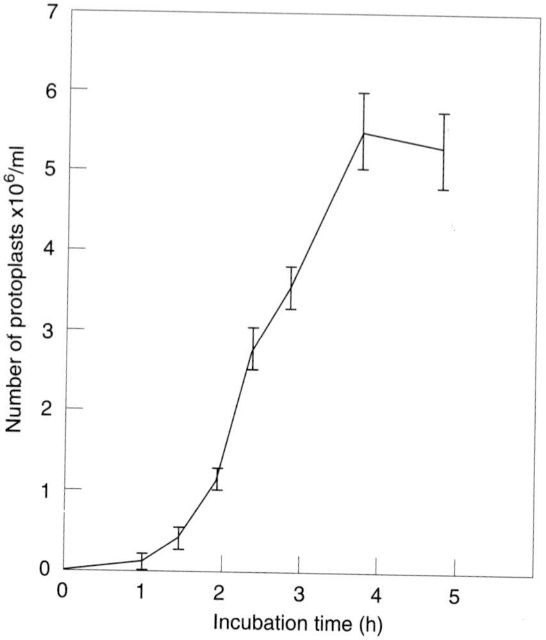

Fig. 3. Release of protoplasts as a function of time under standard conditions. bar markers represent standard variation from the mean values, (Reproduced from Kelkar *et al.* 1990).

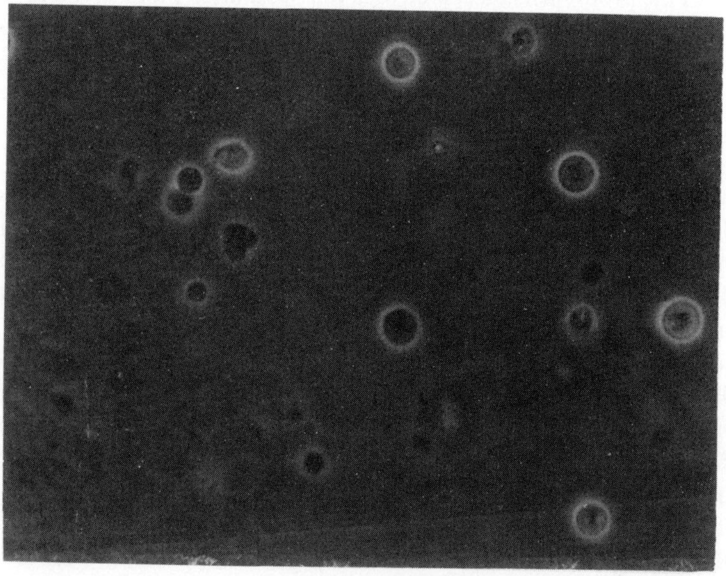

Plate 1. Protoplast released from *Bipolaris oryzae* (Annamalai and Lalithakumari, 1993).

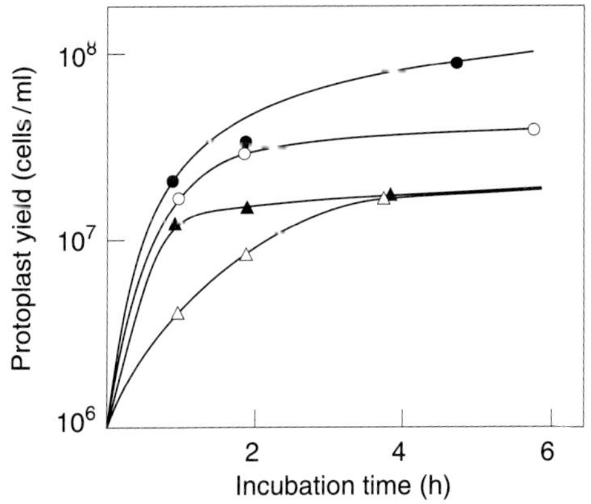

Fig. 4a. Effect of incubation time on the release of protoplast in *Robillarda* sp.
Source: *Kuwabara et al.,* 1989. Enzy. Microb. Technol.

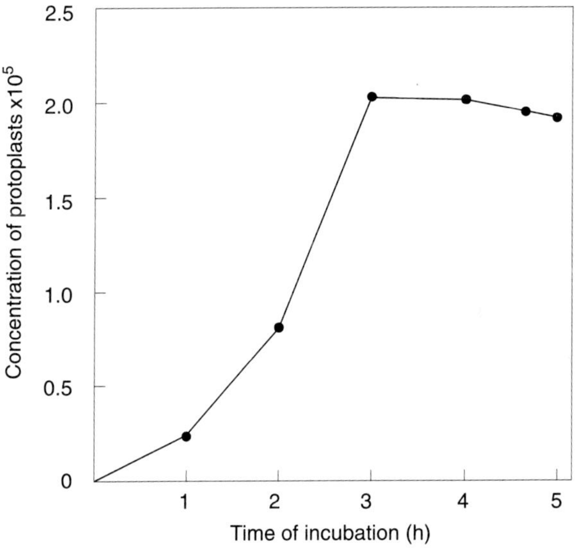

Fig. 4b. Incubation time on the release of protoplasts, from mycelium of *D. oryzae*.
Source: *Annamalai 1989.* Ph.D. Thesis, University of Madras, India, 97-98.

Maximum release of protoplasts was obtained after 3 h of incubation and beyond 3 h did not increase the yield of protoplasts (Vijayapalani, 1995). Mrinalini (1997) has reported (Plate 2a and b) the release of small sized protoplasts in the early hours of incubation and steady medium sized protoplasts after 3 h of incubation with Novozym 234 in *T. harzianum* and 1.5 h in *T. longibrachiatum* (Table 5).

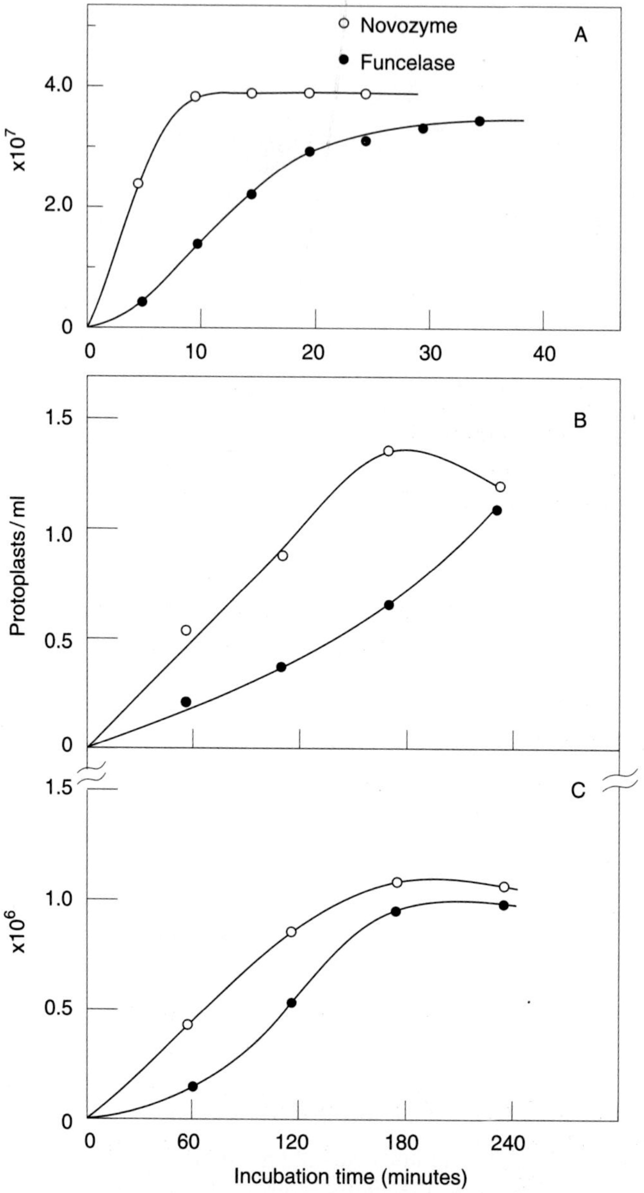

Fig. 5. Liberation of protoplasts with time

(a) *E. ashbyii* (b) *T. reesei* (c) *P. chrysogenum.*

Source: Lakshmi and Chandra, 1993, Enzyme. Microb. Technol. 15, 699-702.

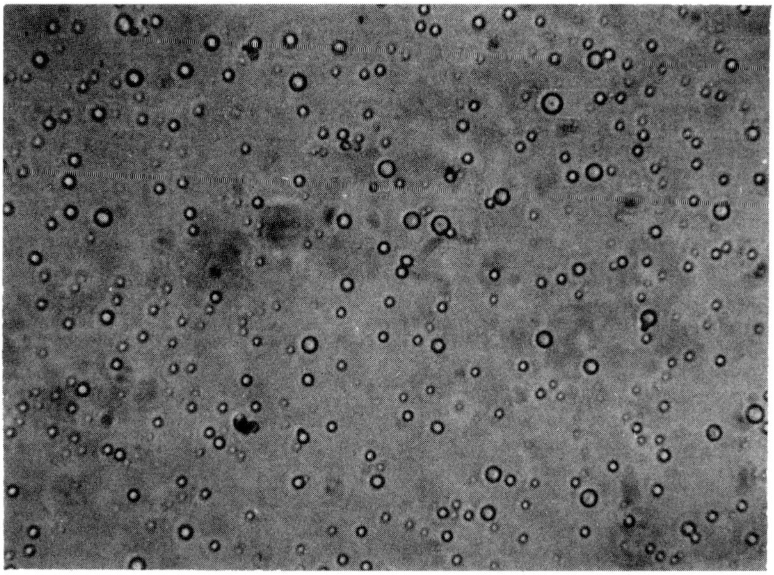

Plate 1a. *V. inaequalis* Protoplasts (× 20).
Source: Vijayapalani, 1995, Ph.D. Thesis, University of Madras, India, 209-213.

Table 4 Effect of time on the release of protoplasts of *Venturia inaequalis*

Incubation time (h)	No. of protoplasts ($\times 10^3$)/mg fresh wt of mycelium
0.5	0
1.0	0
1.5	1.2
2.0	1.5
2.5	1.6
3.0	3.1
3.5	3.1
4.0	3.1

Source: Vijayapalani, 1995. Ph. D. Thesis, Univ. of Madras, 97-225.

Table 5 Effect of incubation time on the release of protoplasts ($\times 10^6$) in *Trichoderma* spp.

Time of incubation (min.)	*T. harzianum*	*T. longibrachiatum*
60	1.2	2.5
90	2.0	3.0
180	3.5	3.0

Source: Mrinalini, 1997. Ph. D. Thesis, Univ. of Madras, India 81-88.

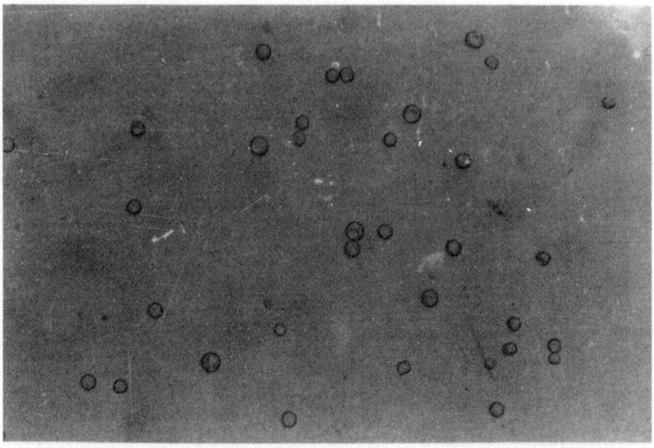

(a)

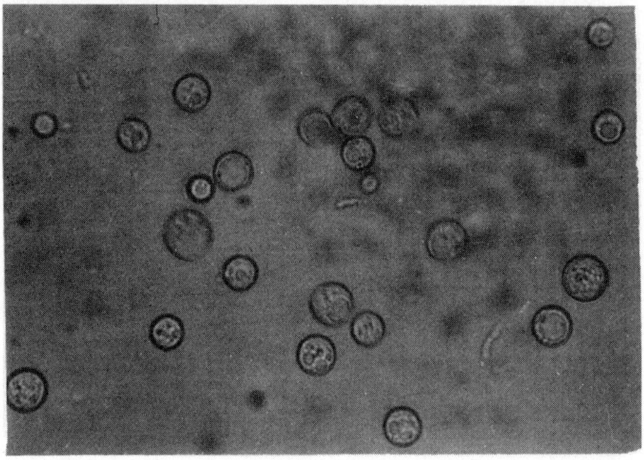

(b)

Plate 2 (a) Very small protoplasm (in the early hours of release). (b) Spherical, medium sized protoplasts.
Source: Mrinalini, 1997. Ph.D. Thesis, University of Madras, India, 81-88.

1.2 FACTORS AFFECTING THE RELEASE OF PROTOPLASTS

Besides the combinations of lytic enzymes, concentration and incubation time, other factors like the age of culture, different (spore and hyphal) stages of the fungus, culture medium, pre-treatment of mycelium, osmotic stabilizers and hydrogen ion concentration also affect the maximum release of the protoplasts.

a. Age of the Culture

In filamentous fungi the life cycle starts with the germination of conidium or ascospore or basidiospore or zoospore or sclerotium depending on the fungal species. The young conidium on germination produces primary hypha which later on branches out to a complex network of mycelium conspicuous on solid medium as colonies.

Mycelial age is a very important factor in protoplasts release. Normally, fungal protoplasts are prepared from young mycelium which is more sensitive to the lytic action of digestive enzymes than old mycelium (Villanueva and Acha, 1971). Release of protoplasts through pores is seen in young mycelia of *Trichothecium roseum* after 3h of Novozym 234 heatment (Plate 2c).

Isolated cell walls of *Geotrichum candidum* obtained from cells at different culture age varied in their susceptibility to lysis with snail digestive juice (Sagara and Tokoshima, 1969). Cell wall of cultures in the early and mid-exponential phase of growth was more susceptible to lysis than cell wall derived from old cultures of the fungus (Peberdy, 1979).

The need for actively growing young hyphae for the production of protoplasts has been reported by Gwinn and Daub (1988) in *Neurospora* and *Cercospora* sp. Kuwabara *et al.* (1989) have clearly shown the marked effect of the age of the cultures on protoplast release in *Robillarda* sp. Y20 which is a filamentous fungus producing cellulases.

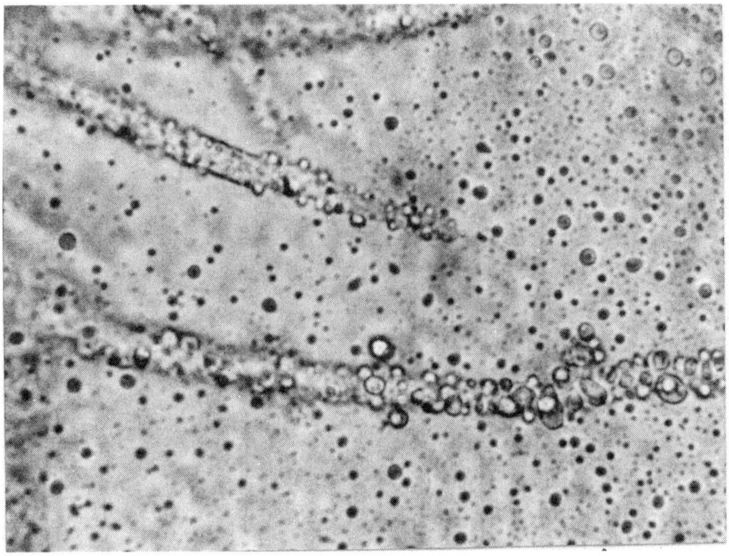

Plate 2c. Release of protoplats from young mycelium of *Trichothecium roseum*.
Source: D. Lalithakumari, 1999 (unpublished)

Mycelia of 24-144 h old were tested for protoplast release (Fig. 6) of *S. rolfsii* (Kelkar *et al.*, 1990). Best results were obtained from 24 h old mycelia and there was a sharp decrease in the yield of protoplasts with increase in the age of the mycelia. To determine the optimum age of the mycelium for release of protoplast from *Dreschlera oryzae*, mycelial cultures of different ages were used (Annamalai and Lalithakumari, 1991). Nearly 100% protoplasts transformation appeared from 36 h old mycelia. Older cultures released very few protoplasts illustrating the importance of physiological age of the mycelium for isolation of protoplasts.

In *Venturia inaequalis* (Fig. 7) young mycelia (24 h) was best suitable for protoplast release and the old mycelia (48, 72 and 96 h) yielded less number of protoplasts (Vijayapalani, 1995).

The yield of protoplasts per mg fresh weight was much higher from young mycelium than from the older ones. The maximum and rapid release from 24 h old mycelium as compared to that from 48 to 72 h old mycelium, suggested that the physiologically active (log phase) young mycelia are more sensitive to enzyme mixture than the mycelium from the stationary phase of growth.

The optimum yield of protoplast was obtained using 15-18 h old unbranched mycelia in *T. viride* (Tamova *et al.*, 1993). In *T. harzianum* and

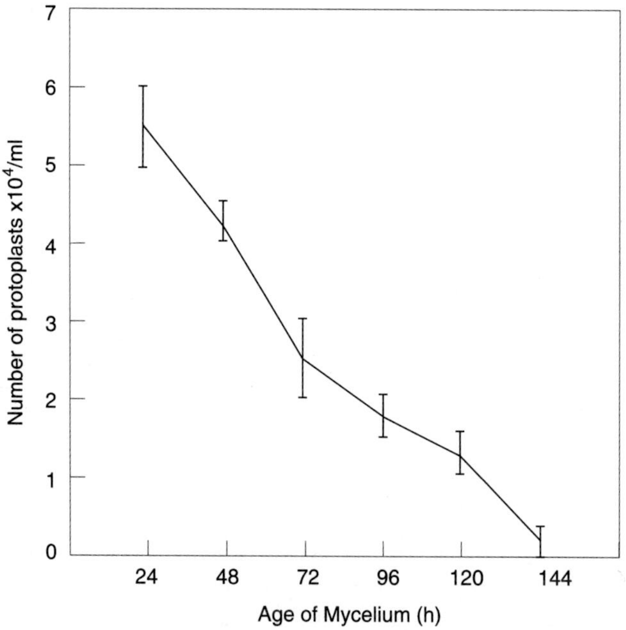

Fig. 6. Protoplast formation in relation to mycelial age under standard conditions. Bar markers represents standard variation from the mean value.
Source: Reproduced from Kelkar *et al.*, 1990.

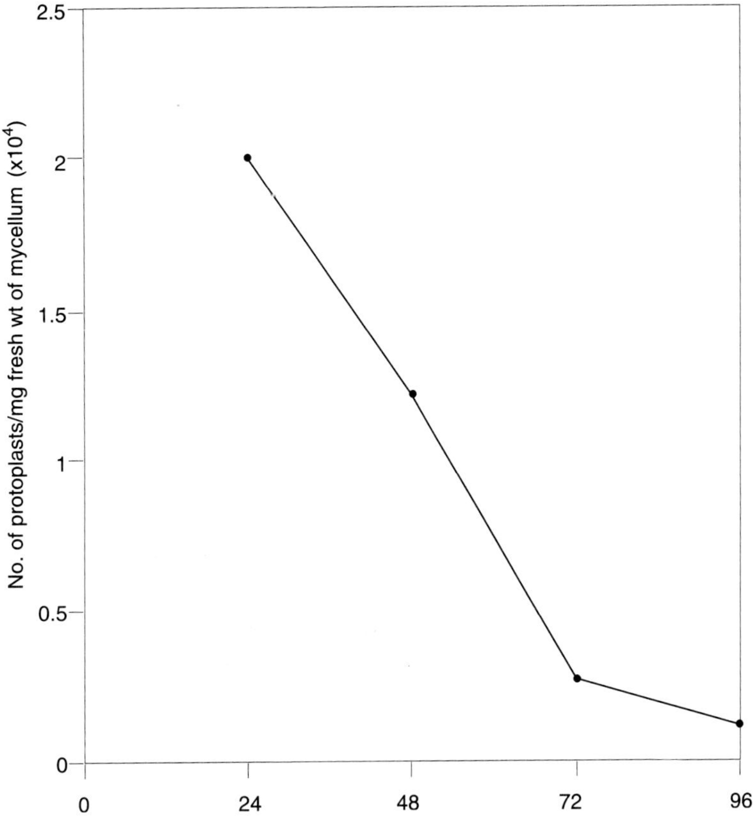

Fig. 7. Age of the mycelium on the release of protoplasts from mycelium of *V. inaequalis*
Source: Vijayapalani, 1995. Ph.D. Thesis, University of Madras, India, 97-225.

T. longibrachiatum, 18 and 24 h culture yielded maximum protoplasts (Fig. 8) while old cultures yielded extraordinarily large protoplasts with poor regeneration capacity (Mrinalini and Lalithakumari, 1998; Karpagam, 1994).

In *Colletotrichum capsici* also 18 h old culture yielded the maximum number of protoplasts (Kalpana, 1995).

Walls from cultures in the early and mid-exponential phase of growth were more susceptible to enzyme digestion than walls derived from older cultures of the fungus. Perberdy *et al.* (1976) found that protoplast yield from *Aspergillus flavus* was highest when cultures in the exponential phase of growth were used, suggesting a possible involvement of endogenous lytic enzymes in the process. Culture age also affects protoplast release in *Saccharomyces cerevisiae.* Young cells from exponential cultures are readily

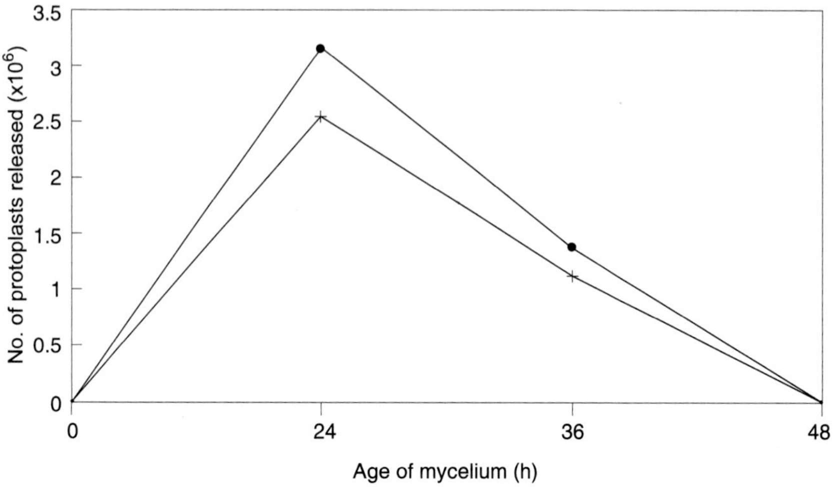

Fig. 8. Effect of the age of mycelium on the release of protoplasts. Th. *T. harzianum:* *Tl. T. longibrachiatum*
Source: *Mrinalini, 1997*. Ph. D. Thesis, University of Madras, 66-102.

converted to protoplasts. Brown (1971) demonstrated that resistance developed rapidly in the transitional period from exponential to the stationary phase of growth. Release of protoplast was maximum from young mycelium (2 day old) than comparatively older cultures in *Volvariella diplasia* (Khanna, *et al.*, 1991). Rapid release of protoplasts was recorded in yeasts (24 h) culture, with Novozym 234 (Plate 3a).

b. Stages of the Fungus

The growth stage of the fungus is another important factor for the successful release of protoplasts. The release efficiency depends on the chemical composition and rigidity of the cell wall, at different stages during the life cycle which is not the same. The composition of cell walls of fungal species varies at different stages of growth. In the same fungus, the substances present in the young hyphae may disappear almost completely as the hyphaspe become older or during the production of asexual and sexual spores or resting spores. The chemical composition is different in different stages like growing mycelium, conidia, chlamydospores, sclerotia etc.

Hyphal state of a fungus is more suitable for the release of maximum protoplasts required for biological studies. However, the culture conditions for largescale production of mycelia need standardization. The germlings or young mycelium is the best for the release of protoplasts. The release of protoplasts from young mycelia is easier owing to their high susceptibility to various lytic enzymes.

i. Spore protoplasts

The protoplast released from hyphae are heterogeneous because they originate from different parts of the mycelium. In contrast, conidial protoplast might to some extent show homogeneity in the number of nuclei per protoplast. Though there are many reports on hyphal protoplast, only a few studies exist on the isolation of protoplasts from spores, probably because of the lack of suitable lytic enzymes. Protoplasts have been released from conidia of *Neurospora crassa* (Bachmann and Bonner, 1959; Manocha, 1968), *Trichothecium roseum* (Garcia Acha *et al.*, 1963) and *Aspergillus favus* (Garcia *et al.*, 1966). In *Fusarium culmorum*, protoplasts from macroconidia have been veleased. Asai *et al.* (1986); Lalithakumari and Annamalai (1993) reported spore protoplast in *P. oryzae* (Plate 3a & b) and in large quantity by using commercial enzymes (Table 6). Zymolase, Driselase and β-glucuronidase were found to be able to release protoplasts from spores. Mixture of enzymes produced more protoplasts. *Alternaria* sp. conidia released protoplasts (Plate 3b) when treated with Novozym 234.

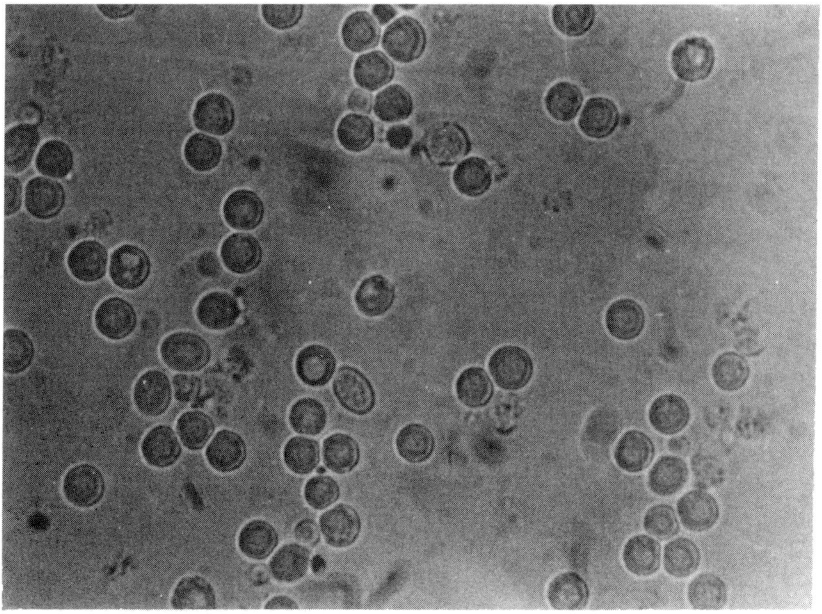

Plate 3a Release of protoplast from young yeast cultures (24 hrs old.)
Source: D. Lalithakumari 1999 (unpublished)

Table 6 Isolation of protoplasts from spores of *Pyricularia oryzae* by commercial enzymes [a]

Enzymes	Yield of protoplasts [b]
Zymolyase	++
Driselase	++
β-glucuronidase	+
Cellulase	–
Helicase	–
Macerozyme	–
Pectriase	–
Zymolyase and β-glucuronidase	+++
Driselase and β-glucuronidase	+++
Zymolase and Driselase	+++

[a] Spores (10^6/ml) were incubated at 37° C for 6 h each enzyme at the concentration of 1% in 0.6 M KCl at pH 5.5.

[b] Number of isolated protoplast in 1 ml of enzyme solution - + + +: > 3×10^5;+ +;3×10^5; –; +; 1×10^5; –: 0.

(Asai et al., 1986).

Spores are a very useful source for protoplast isolation from many filamentous fungi, especially in species where they are uninucleate. Spore protoplasts are more homogenous in their organelle composition and therefore in their behaviour. In general, the spore wall is more resistant to lysis and attempts to isolate protoplasts have been successful in only a few instances (Fig. 9). They were obtained either following long periods of lytic digestion (Laborda *et al.*, 1974; Lalithakumari & Annamalai, 1991) or by using spores produced in shaken liquid culture (Moore and Peberdy, 1976a)

ii. Protoplasts from batch cultures

Protoplasts isolated from batch cultures of yeast or filamentous fungi, by definition, must be heterogeneous with regard to their physiology and biochemistry. The yeast culture, being asynchronous, is comprised of cells at different stages of cell cycle, and this will be reflected in the protoplasts

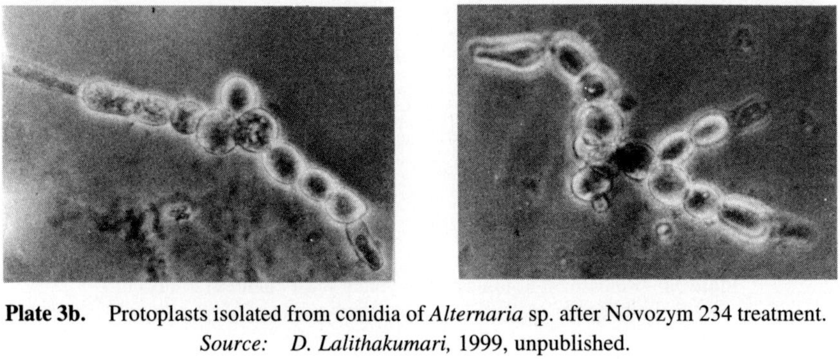

Plate 3b. Protoplasts isolated from conidia of *Alternaria* sp. after Novozym 234 treatment. *Source: D. Lalithakumari,* 1999, unpublished.

derived from them. In the filamentous fungi, the heterogeneity arise for a different reason, namely as a reflection of the organelle and biochemical differentiation that is found in the fungal hypha (Grove, 1978). The occurrence of a duplication cycle in the terminal segment of hyphae has significant and interesting implications in relation to protoplasts release (Trinci, 1978). At the population level the physiological status of the culture is a major factor in determining protoplast yield.

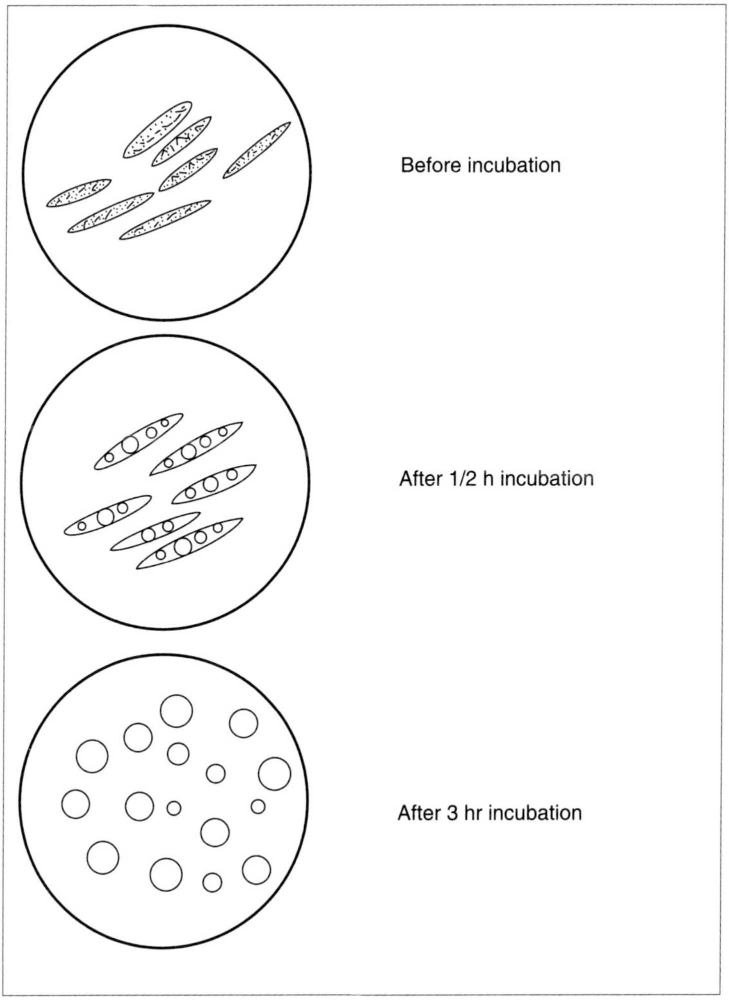

Fig. 9. Schematic representation of formation of protoplasts from conidia of *Bipolaris oryzae*.

c. Effect of Concentration of Fungal Mycelium

The number of protoplasts released is proportional to the concentration of mycelium. The concentration of mycelia vs lytic enzyme activity and incubation is to be optimized for individual organisms. Mycelial wet weight and number of protoplasts released can easily be monitored. However, the liberated protoplasts, after 3-5 h incubation in suspension at high mycelial concentrations, were smaller than those in suspension at low mycelial concentrations and protoplast of high mycelial concentrates did not contain any vacuoles. These small protoplasts were probably released from hyphal tips, which were clearly more abundantly present at higher mycelial concentration and digested preferentially to older mycelial parts (Anne, 1977). The number of protoplasts released were maximum when lesser concentrations (100 mg wet wt) of mycelia were used. In higher concentrations (200 and 300 mg wet wt) there was a decrease in the number of protoplasts released (Table 7) in *Trichoderma harzianum* and *T. longibrachiatum*.

Table 7 Effect of concentration of mycelia on the release of protoplasts in *Trichoderma* spp.

Strains of *Trichoderma*	Mg fresh wt. of mycelia	No. of protoplasts released ($\times 10^6$)
T. harzianum	100	3.1
	200	2.5
	300	1.3
T. longibrachiatum	100	2.5
	200	2.0
	300	1.2

Source: Mrinalini, 1997. Ph. D. Thesis, Univ. of Madras 81-83.

d. Culture medium

Culture medium is also an important pre-requisite for protoplast release. Musilkova and Fencl (1968) showed the dramatic effect of the culture of *Aspergillus niger* on a defined medium and on a complex medium. Yields obtained from mycelia grown on the defined medium were impressively high. Protoplasts production from *Saccharomyces pombe* was enhanced from cells grown in the presence of 2-deoxy-D-glucose (Birnboim, 1971; Foury and Goffeau, 1973).

e. Pre-treatment of Mycelium

The cells have also been given some pre-treatment that improve the protoplast yield. Thiol compounds have been used extensively with yeasts (Kovac *et al.*, 1968; Darling *et. al.*, 1969; Kuo and Lampen, 1971; Foury and Goffeau,

1973; Houssett *et al.*, 1975; Yamamura *et al.*, 1975) and some filamentous fungi, including *Cephalosporium acremonium* (Fawcett *et al.*, 1973), *Histoplasma capsulatum* Berliner and Reca, 1969) and *G candidum* (Dooijewaard-Kloosteriziel *et al.*, 1973). Thc effect of these compounds in enhancing protoplast release has been related to the reduction of disulfide bonds in wall proteins, thus opening up the molecules and allowing penetration of the lytic enzymes (Anderson and Millbank, 1966). Another treatment shown to improve protoplasts involved the use of Triton X-100 with *Pythium* (Sietsma and De Boer, 1973). The enhancing effect was presumed to result from the removal of a lipid layer. This idea was further supported by experiments in which lipase was found to be an adjuvant in protoplast isolation from the fungus (Eveleigh *et al.*, 1968).

Pre-treatment of mycelium with reducing agents such as β-mercaptoethanol and dithiothreitol is necessary for a few fungi (Tyagi *et al.*, 1981). Pre-treatment of mycelium or cells with a thiol compound prior to digestion with lytic enzymes has generally been found to be necessary with *Acremonium chrysogenum* and *Saccharomyces cerevisiae*. *A. chrysogenum* mycelium was pre-treated for 1 h in citrate phosphate buffer (pH 7.3) containing 0.01 M dithiothreitol and *S. cerevisiae* cells were pre-treated with 0.7% (v/v) mercaptoethanol in buffer (pH 7.8) for 5 min (Hamlyn *et al.*, 1981). Harris (1982) showed that protoplasts were produced from *Fusarium* sp. only after the treatment of mycelium with solution containing β-mercaptoethanol, EDTA and cysteine. Thiol compounds may destabilize the disulphide linkages in the cell wall thereby allowing easier accessibility for cell wall lytic enzymes. Pre-incubation in EDTA might also cause the same effect to provide proper conditions to form protoplasts (Pfeifer and Khachatourians, 1987). Pre-treatment is not required universally. Pre-treatment of the *S. rolfsii* mycelium with dithiotheritol (5 mM) and β-mercaptoethanol (1 l/ml) when used singly and in combination with EDTA (50 mM) did not show any appreciable change either in the number of protoplasts released or in the time of release (Kelkar *et al.*, 1990). Vijayapalani (1995) reported the importance of pre-treatment with β-mercaptoethanol for efficient release of protoplasts in *V. inaequalis*.

f. Osmotic Stabilizers

An osmotic stabilizer is essential to provide osmotic support to the protoplasts following the removal of the cell wall. An extensive range of inorganic salts, sugars and sugar alcohols have been successfully used as osmotic stabilizers.

Due to lack of external protection, protoplasts would immediately lyse during the process of release from mother cells, unless osmotic concentration is favourably adjusted (Peberdy, 1979; Davis, 1985). A wide range of osmotic stabilizers have been used to achieve maximum number of protoplasts from fungi but n universal stabilizer has been reported for all fungi. The type and concentration of osmotic stabilizer influence both yield and stability of protoplasts. However, inorganic salts are effective stabilizers in several species. The concentration of solutes varies largely and this phenomenon can be correlated to some extent with difference in the internal osmotic pressure of different species. The nature and concentration of the stabilizer is a critical factor in protoplast release.

Best results on protoplast release in *Fusarium culmorum* were obtained with 0.8 M NH_4Cl, mannitol, NaCl or KCl and rhamnose. Na_2CO_3, NaH_2PO_4 and $(NH_4)_2SO_4$ at an optimum pH of 6-8 did not yield favourable results (Lopez Belmonte *et al*., 1966).

In general, inorganic salts are effective with filamentous fungi and sugars or sugar alcohols are more effective with yeasts. The virtues of particular stabilizers are understood only in an empirical sense, and differences in effectiveness must relate to as yet unknown factors in the uptake and utilization of the particular compounds. One of the most interesting observations of osmotic stabiliizers was made by De Vries and Wessels (1972), who identified a distinct property exhibited by 0.6 M magnesium sulphate when used with filamentous fungi. Two effects were observed. In the first type, the mycelium became extensively fragmented in the early period of lytic digestion. A large proportion of the protoplasts released had large vacuoles which accounted for the greater part of the protoplast volume. The vacuoles, yielded an added benefit, because after centrifugation, the protoplasts that floated could easily be removed to give preparations uncontaminated with mycelial debris. The second type of protoplast was sedimented under gravity. $MgSO_4$ induced bouyancy was found in protoplasts of *Aspergillus nidulans* (Peberdy and Isaac, 1976). Large protoplasts (Plate 4 a and b) are released with very poor regeneration when $MgSO_4$ is used as osmotic stabilizer due to vacuolization (Revathi, 1993) in *V. inaequalis*. Further Vijayapalani (1995) has reported shrinkage and lysis of protoplasts of Ambri strain of *V. inaequalis* even though they were larger in size with $MgSO_4$ as osmotic stabilizer.

In *Fulvia fulva*, mycelium from liquid cultures of 24-48 h incubated with Novozym 234 in buffered 1.0 M $MgSO_4$ gave the best conditions for protoplast release. Protoplasts released in $MgSO_4$ were large and highly vacuolated (Harling *et al*., 1988). The use of $MgSO_4$ and KCl instead of organic compounds also has the advantage of preventing bacterial growth during incubation (Gascon *et al*., 1965).

Sugars affect the synthesis of cell wall and also autolysis. Thereore, it is ncessary to determine the optimal concentration of sugar for each isolate. Davis (1985) suggested solutions of inorganic salts as the stabilizers for

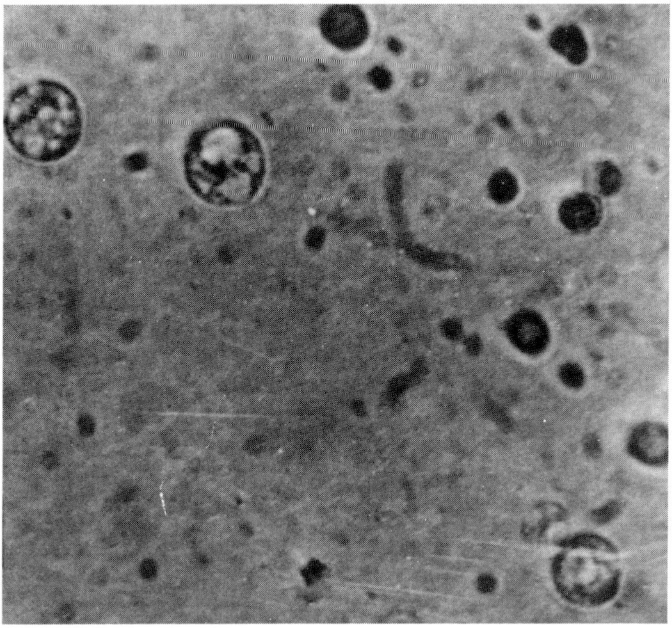

Plate 4a. Large protoplasts ($MgSO_4$ as osmotic stabilizer) of *V. inaequalis*.
Source: *Revathi, 1993*. Ph.D. Thesis, University of Madras, India, 71-155.

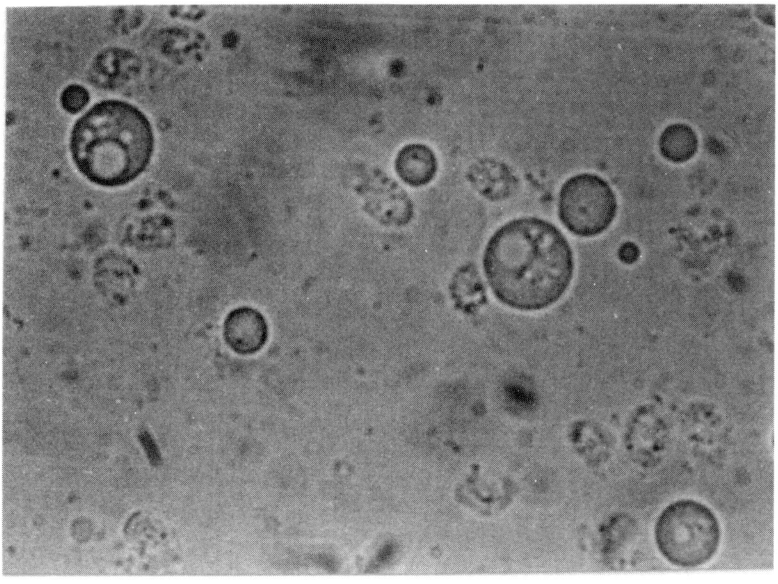

Plate 4b. Large protoplasts ($MgSO_4$ as osmotic stabilizers) of yeasts.
Source: *D. Lalithakumari, 1999,* Unpublished.

protoplasts from filamentous fungi. Levitin *et al.* (1984) and Turgeon *et al.* (1985) suggested KCl as the best stabilizer for *Drechslera tereus* and *Cochliobolus heterostrophus*. Osmotic stabilizers play an important role in the release and maintenance of the integrity of the protoplast (Hocart *et al.*, 1987; Mukherjee and Sengupta, 1988).

$MgSO_4$ and to a lesser extent, mannitol and sorbitol are effective stabilizers, while no protoplasts were produced in presence of NaCl and KCl in mycorrhizal fungi, *Hebeloma cylindrosporum.* Highest yield of protoplast was obtained using 2 day-old mycleia with 0.7 M $MgSO_4$ as osmotic stabilizer (Hebraud and Fevre, 1987). In general, the osmotic stabilizer affects the relative density of protoplasts. In NaCl, KCl, mannitol and sorbitol, the protoplast sedimented under gravity and in sucrose, the protoplasts had a similar density to that of the solute and remained in suspension.

The molarity of the osmotic stabilizer also plays an important role in the stability of protoplast and regeneration frequency. Regeneration frequency of *S. rolfsii* protoplasts was maximum (90-95%) when sucrose (0.6 M) was used as osmotic stabilizer though lower yields of protoplasts were obtained with sucrose and sorbitol (Kelkar *et al.*, 1990). Favourable stabilizers for protoplast isolation are 0.4 M NaCl and 0.7 M mannitol in case of *T. harzianum* (Tashpulatov *et al.*, 1991), 0.7 M KCl in phosphate buffer for *T. viride (*Tamova *et al.*, 1993), 0.6 M sucrose-0.8 M mannitol for *Bipolaris oryzae* (Annamalai and Lalithakumari, 1991). Four osmotic stabilizers were tested to evaluate their efficacy in releasing protoplasts from *B. oryzae* mycelium. Large numbers were obtained with 0.6 and 0.8 M sucrose-mannitol than with other osmotica (Table 8).

Table 8 Effect of osmotic stabilizers on the yield of protoplasts from mycelium of *B. oryzae*

Stabilizer	Molarity	No. of protoplasts ($\times 10^5$)/mg Mycelium
Mannitol	0.6	0.5
	0.8	0.5
Sucrose	0.6	0.5
	0.8	0.5
KCl	0.6	0.2
	0.8	–
$MgSO_4$	0.6	–
	0.8	–
Sucrose-mannitol	0.6	2.0
	0.8	1.2

Source: Annamalai and Lalithakumari, 1991. J. Plant disease and protection, 98, 197-204.

The yield was increased to a maximum of 2.0×10^5 when sucrose-mannitol was used, while it was 0.5×10^5 in 0.6 and 0.8 M mannitol. The yield was highly reduced, when $MgSO_4$ (0.6 and 0.8 M) and KCl (0.6 and 0.8 M) were used as osmotic stabilizers.

Among the seven osmotic stabilizers (Table 9) used for the release of protoplasts from the mycelium of *V. inaequalis* (Revathi and Lalithakumari, 1992), sucrose, sorbitol, mannitol, $MgSO_4$, NaCl and KCl were not suitable as the number of protoplasts released were few. Vijayapalani (1995) also reported a combination of sorbitol-sucrose (0.6–M) for maximum number of protoplast release (Table 10) in *V. inaequalis* (Ambri strain).

Table 9 Comparison of osmotic stabilizers on the release of protoplasts from *V. inaequalis*

Osmotica	Molarity (Mol/L)	No. of protoplasts ($\times 10^4$)
$MgSO_4$	0.6	0.3
NaCl	0.6	0.2
Sucrose	0.6	0.6
Mannitol	0.6	0.2
Sorbitol	0.6	0.6
KCl	0.6	0.2
Sorbitol-sucrose	0.6	2.0

Source: Revathi and Lalithakumari, 1992, Z. Pflangenkrankh. Pflangensch. 100: 211-219

Even though the protoplast size was larger when $MgSO_4$ was used as a stabilizer, the released protoplasts exhibited shrinkage and further incubation led to lysis of protoplast (Vijayapalani, 1995). The size of protoplasts varied from 2-8 μ. The absence of cell wall in the isolated protoplasts was indicated by the observations that they were not stained with calcoflour white. But in undigested mycelial bits the cell wall brightly fluoresced with blue tinge upon staining with this dye.

Protoplasts were released from *Trichoderma* spp. (Table 11) and *R. solani* using 0.6 M KCl (Elavarasan, 1996) and 0.6 M NaCl (Table 12) in *Colletotrichum capsici* (Kalpana, 1995).

In *Volvariella diplasia*, 0.6 M concentration of KCl was the best osmotic stabilizer for the release of protoplast (Khanna *et al.*, 1991) $MgSO_4$ $7H_2O$ at 0.4 M concentration has been reported to be a good osmotic stabilizer for the release of protoplast from *Volvariella volvacea* (Chang *et al.*, 1985; and KCl proved the best for continuous edodes (lijima and Yangi, 1986).

Table 10 Efficiency of osmotica on the release of protoplasts from *V. inaequalis* (Ambri)

Osmoticum/osmotica	No. of protoplasts ($\times 10^3$) mg fresh wt of mycelium at different molarity		
	0.4	0.6	0.8
Sucrose	1.3	2.0	2.0
Sorbitol	1.3	2.2	2.1
Mannitol	1.1	1.5	2.0
Magnesium sulphate	0.2	0.9	1.2
Magnesium chloride	1.0	0.3	0.1
Sodium chloride	0.3	0.3	0.5
Potassium chloride	0.1	0.1	1.5
Potassium acetate	0	0	0.1
Sodium acetate	0	0	0.2
Amonium chloride	0	0	0.3
Ammonium sulphate	0	0.1	0.1
Sucrose + Sorbitol	1.7	3.1	2.7

Source: Vijayapalani, 1995. Ph. D. Thesis, Univ. of Madras, India 209-213.

Table 11 Effect of various osmotic stabilizers on the release of protoplasts from *Trichoderma* spp.

Osmotic stabilizer	Molarity (mol/L)	Strains of *Trichoderma*	No. of protoplasts released ($\times 10^6$)
KCl	0.6	Th	3.1
		Tl	2.5
NH_4Cl	0.6	Th	2.0
		Tl	1.8
Sucrose & Sorbitol	0.6	Th	2.1
	0 .6	Tl	1.8
$MgSO_4$ &	1.4	Th	2.1
Sodium citrate	0.05	Tl	1.0

Th: T. harzianum; Tl: T. longibrachiatum
Source: Mrinalini, 1997. Ph. D. Thesis, Univ. of Madras, India 81-83.

Table 12 Effect of osmotica on the release of protoplasts of *C. capsici*

Osmotica	Molarity (mol/L)	No. of protoplasts released ($\times 10^4$)
NaCl	0.8	6
NaCl	0.6	12
KCl	0.8	3
KCl	0.6	5

Source: Kalpana, 1995. M. Phil Thesis, Univ. of Madras, India, 27-29.

g. Hydrogen Ion Concentration

The pH of the osmotic stabilizer is an important factor for release since the pH determines the activity of the different hydrolytic enzymes present in the enzyme preparation. Generally, protoplasts are obtained at pH ranging from 4.9 to 6.2. The maximum number of protoplasts were observed around pH 5.8 in *Schizophyllum commune* (De Vries and Wessels, 1972). In the case of the basidiomycete *Polystictus versicolor* (Strunk, 1969) pH 7 was best for protoplast release. In *T. viride* (Tamova *et al.*, 1993), pH 6.8, in 0.6 M KCl was optimal in *T. reesei* (Kumari and Panda, 1993). In *P. chrysogenum* the mycelium was not digested below pH 4.0 and above pH 8.0.

Protoplasts lysed soon after they emerged. Irrespective of the effective preparations, the maximum number of protoplasts was observed between pH 5.0-6.0 (Anne, 1977) (Fig. 10). Different levels of pH used to monitor the release of a large number of viable protoplasts (Fig. 11) revealed pH 5.5 as more suitable for both. *T. longibrachiatum* and *T. harzianum* when incubated with Novozym 234 (Mrinalini, 1997).

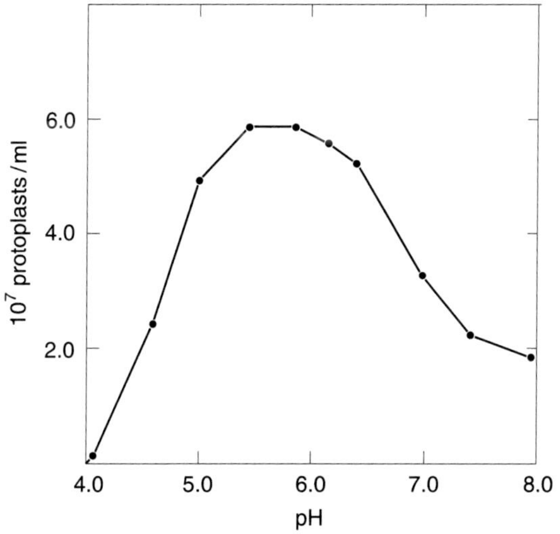

Fig. 10. Influence of pH on the production of protoplasts. Solutions of strepzyme (10 mg/ml) plus TLE (20 mg/ml) in 0.55 M NaCl and 0.1 M phosphate buffer at various pH were compared for production of protoplasts from 16 h old *P. chrysogenum* mycelium (50 mg/ml) in a total volume of 3 ml. Protoplasts were counted after 3 h incubation at 27°C and 150 rpm.
Source: *Anne 1977*, Agriculture, 25, 21-44.

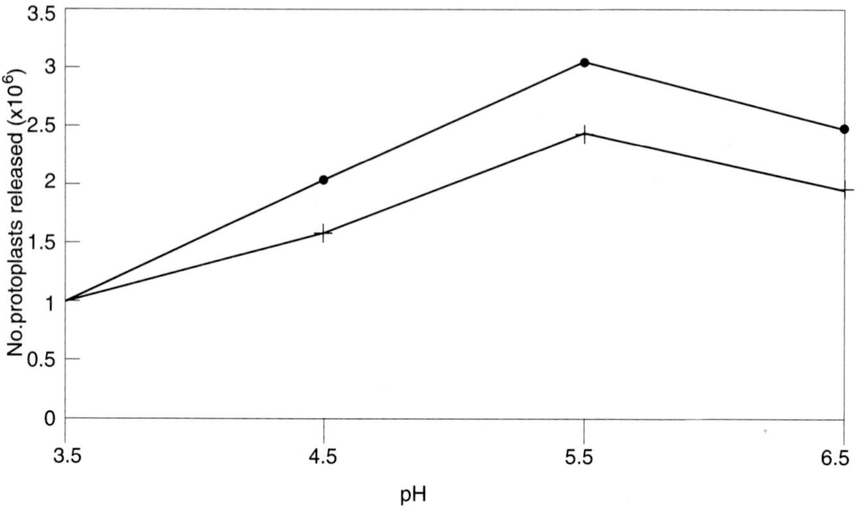

Fig. 11. Influence of pH on the release of protoplasts (Th - *T. harzianum*; Tl - *T. longibrachiatum*)

Source: Mrinalini, 1997. Ph.D. Thesis, University of Madras, India, 81-83.

1.3 MONITORING FUNGAL PROTOPLAST FORMATION (MICROSCOPIC EXAMINATION)

The protoplast formation can continuously be monitored under a microscope. Filamentous fungi release their protoplasts as osmolabile spherical bodies through pores in the cell wall produced by the action of mycolytic enzymes (Fig. 12).

The release of protoplasts from filamentous fungi is the same irrespective of the lytic enzyme used.

The time taken for the release of protoplasts varies with different fungi. In *P. chrysogenum* (Anne, 1977), the hyphal tips became swollen and the protoplast emerged by an extension of cytoplasm at the digested tips or in their close vicinity. After 3-5 h incubation the mycelium was almost completely transformed into the protoplast leaving behind mycelial debris and empty cell walls.

Size of the protoplasts varies in filamentous fungi. Anne's (1977) observations in *P. chrysogenum* could be a guide line. According to Anne's observation, the early protoplasts released in the first hour of incubation were small spherical bodies, non-vacuolated and homogenous in size but after 2 h of incubation protoplasts became large and often contained vacuoles. Similar differences in size between 'early' and 'late' protoplasts were also recorded

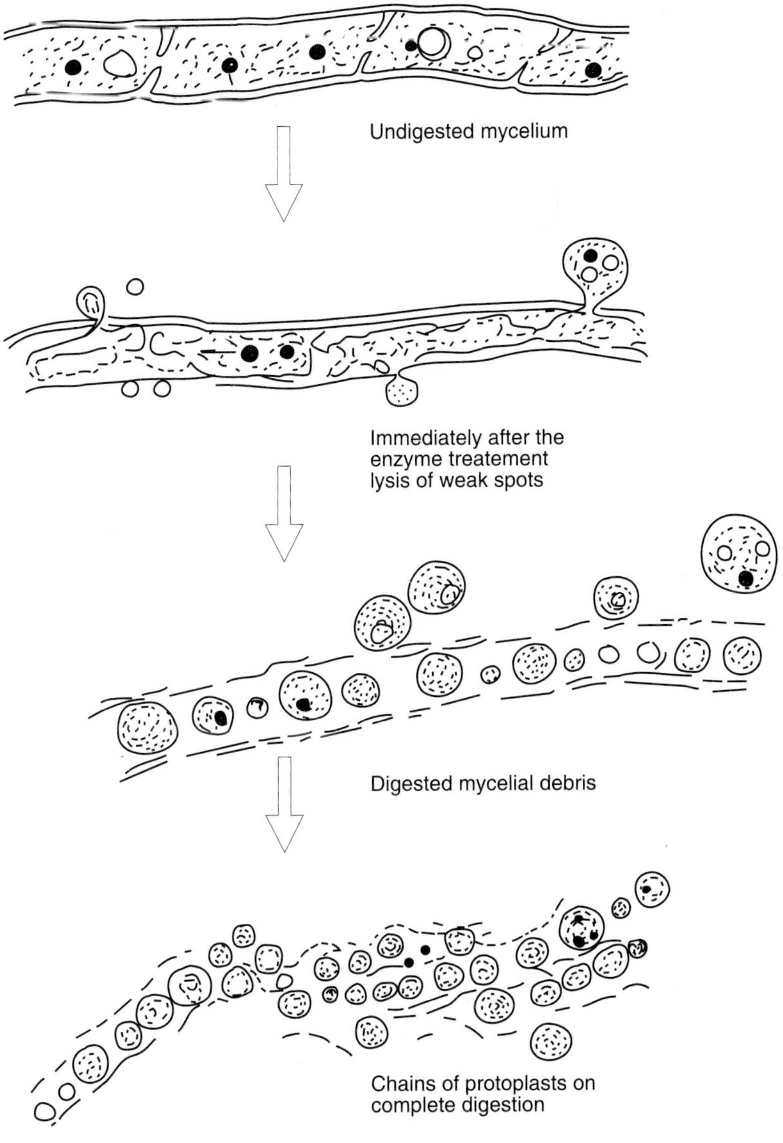

Fig. 12. Schematic representation of the release of protoplasts from filamentous fungi.

by other workers (Gibson and Peberdy, 1972). The protoplasts increased in size from about 4.8 μ after 1 h to 16.8 μ after 48 h and up to 21.9 μ after 96 h incubation. After 3 h, 47.5% of the total number of protoplasts contained 1 nucleus and less than 6% had 4 nuclei, after 24 h in the lytic enzyme solution, less than 3% had 1 nucleus and more than 20% had 4 nuclei.

Protoplasts containing as many as 28 (Plates 5a and b) nuclei were also recorded by Anne (1977).

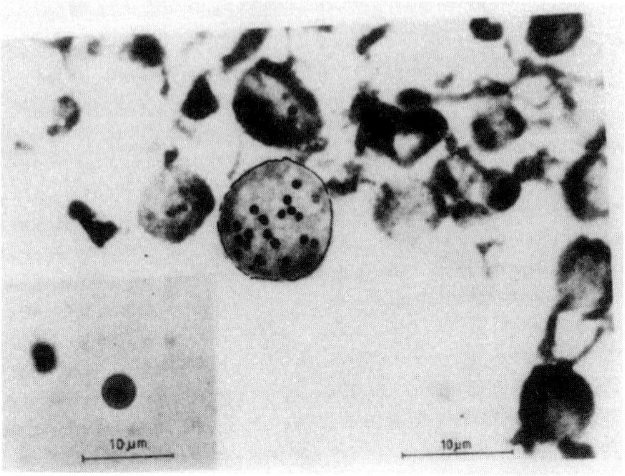

Plate 5a. Giemsa-stained nuclei of *P. chrysogenum* protoplasts. Micrograph of multi-nucleate protoplasts after 24 h incubation in the lytic enzyme solution. The inset shows a mono-nucleate protoplast released after 3 h incubation. Reproduced from Anne, 1997.

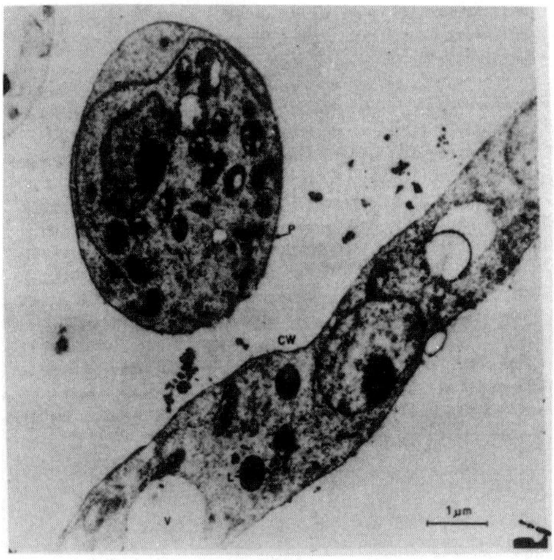

Plate 5b. Electron micrograph of a *P. chrysogenum* protoplast and a non-digested hypha. Key to symbols: CW = cell wall; ER = endoplasmic reticulum; L = lipid storage granule; M = mitochondrion; N = nucleus; NM = nuclear membrane; Nu = nucleolus; P = plasmalemma; S = spectum; T = tonoplast; V = vacuole; Ve = vesicle. Protoplasts were produced during incubation (27°C, 150 rpm, 3 h) of 16 h old mycelium with the strepzyme-TLE cell wall lytic enzyme system in 0.55 M NaCl. Protoplasts were fixed in 1% glutaraldehyde, pH 7.0, for 20 min and post-fixed in 1% OsO_4 for 1 h. Reproduced from Anne, 1977.

In protoplast preparation from *Schizophyllum commune* multinucleate protoplasts were observed after storage in the lytic enzyme solution (Van der Walk and Wessels, 1973). In *V. inaequalis* Ambri strain, Vijayapalani (1995) showed the sequential release of protoplasts after incubation with lytic enzyme. The first observable change in the mycelium on incubation with the lytic enzymes was the swelling of the hyphal tips. It was followed by the fragmentation of mycelium and release of non-vacuolated protoplasts after 1.5 h of incubation. Continued incubation resulted in swelling of the older parts of the hyphae and release of large sized vacuolated protoplasts. A chain of protoplasts (Plate 6a & b) was observed in the incubation medium after the complete digestion of the cell wall of *V. inaequalis. In V. inaequalis* typical release of protoplasts through pores leaving empty cell wall behind is observed.

The protoplasts were spherical, immediately after isolation, although they differed in their diameters. Several of the protoplasts contained granular inclusions and vacuoles. The released protoplasts showed heterogenous structure, because they originate from different parts of the mycelium. In other words, the protoplasts were not uniform but were of different sizes and appearances. The contents of the protoplasts were also not uniform and other internal structures i.e. number of nuclei, mitochondria etc. also varied. Some of the protoplasts had one very large vacuole or several vacuoles.

The number of regeneration tubes is an indication of the number of nuclei in a protoplast. A nucleiless or anucleate protoplast does not regenerate, instead produces bud like structures and disintegrated in the course of incubation (Plate 7).

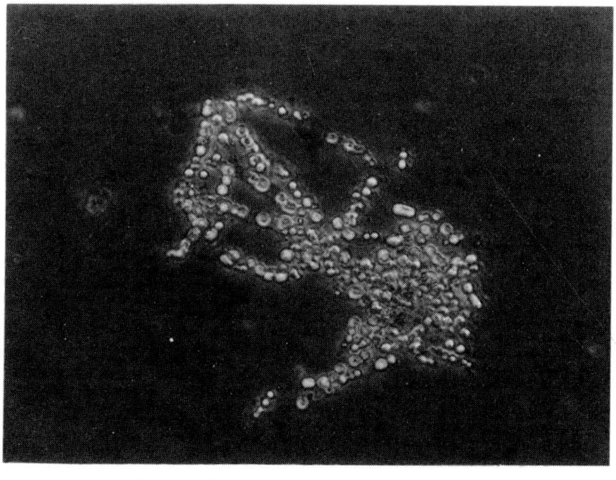

Plate 6a. Release of protoplasts from young hyphae of *V. inaequalis* (320 ×)
Source: Vijayapalani, 1995, Ph.D. Thesis University of Madras, India, 211.

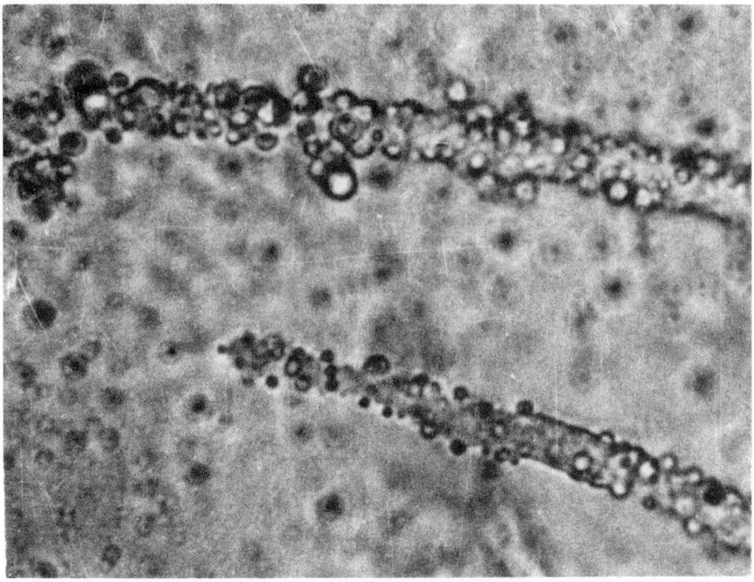

Plate 6b Release of protoplasts of *Trichothecium roseum* through pores of mycelium.
Source: D. Lalithakumari 1999, Unpublished.

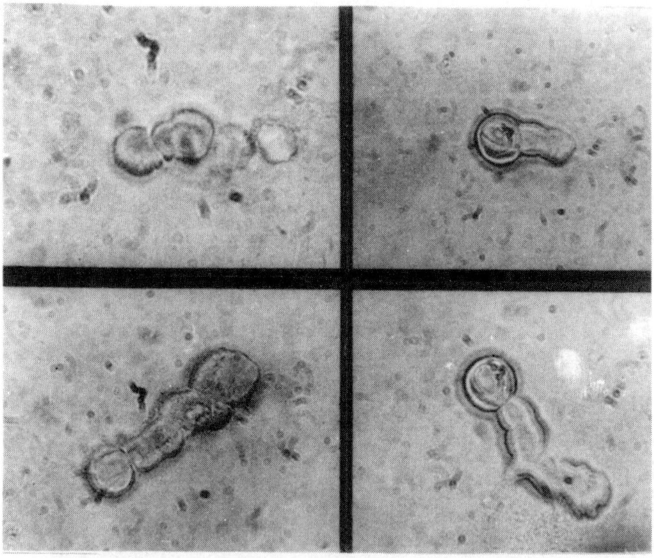

Plate 7. Anucleate protoplast of *V. inaequalis* produce bud-like structures and disintegrate.
Source: Vijayapalani, 1995. Ph.D. Thesis, University of Madras, India 221.

The observation of small protoplasts were always mixed with large sized protoplasts and no such 'early and late' protoplasts were recorded in *V. inaequalis* (Plate 8). This need not be a rule in all fungi and the rapid dissolution of mycelium by lytic enzyme depends on the chemical constitution of the target fungi and mixture of different sizes of protoplasts were always observed in many filamentous fungi. In *B. oryzae* (Annamalai and Lalithakumari, 1991), *P. oryzae* (Aunamalai and Lalithakumari, 1991), *V. inaequalis* (Revathi, 1993; Vijayapalani, 1995) *T. harzianum* and *T. longibrachiatum* (Mrinalini and Lalithakumari, 1993; Karpagam, 1994), *R. solani* (Elavarasan, 1996) and *Colletotrichum capsici* (Kalpana, 1995) variable size of protoplasts were recorded.

In *Fusarium culmorum*, the hyphal content between the two septa liberated into one or more protoplast (Villanueva, 1966). Thus, one protoplasts did not always correspond to a single cell (Lopez-Belmonte *et al.,* 1966). Not all the protoplasts formed from the mycelium are similar in size. Their contents vary, the number of nuclei, vacuoles and cytoplasmic particles may be quite different (Garcia *et al.*, 1966). The number of nuclei varies from 2 to 4 and sometimes many nuclei pass together with cytoplasmic contents through septa as we have seen during he liberation of protoplasts. The number of protoplasts liberated from one hyphal compartment is not constant and varies. Attempts to detect nuclei in protoplasts have sometimes failed. This can often demonstrated by the specific staining techniques used, but the possibility cannot be excluded that they are not observed although they are present in the

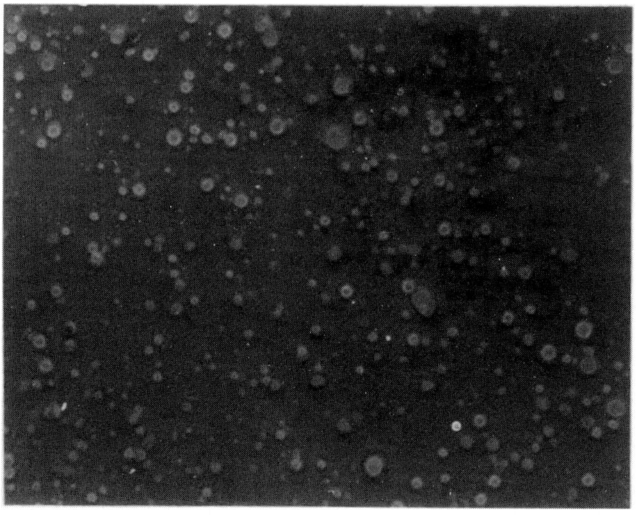

Plate 8. Mixture of different sizes of protoplasts of V. inequalis (260x)
Source: *Vijaypalani, 1995.* Ph.D. Thesis, University of Madras India 212.

body. Most of the protoplasts possess atleast one or two nuclei. The size of the protoplasts is not related to the number of nuclei in it. Based on the observation on the number of germ tubes produced by each protoplast one could predict the number of nuclei in a protoplast as it is definite that an anucleated protoplast does not successfully regenerate.

The number of nuclei could be correlated to the number of germs tubes produced as one, two, three and multinucleated. The importance of nuclear material has always been emphasized in the work of regeneration in plant cells and protozoa.

Only protoplasts possessing at least one nucleus regenerate the cell wall, giving rise to normal mycelial forms.

The degree of fluorescence of the protoplast was an indication of the presence of cell wall material on the membrane (Peberdy and Buckley, 1973). In a suspension of purified protoplasts between 85 to 95% of the total number showed no fluorescence. The other protoplasts showed a weak overall fluorescence or small fluorescing spots at the membrane, indicating that small amounts of cell wall material had been retained.

1.4 PROPERTIES OF THE PROTOPLASTS

The protoplasts were sensitive to osmotic shock. When placed in distilled water, they swelled slightly and burst suddenly. Sometimes one or more optically empty vesicles, apparently vacuoles, remained visible. When placed in hypertonic solution, they shrunk quickly. At the same time one or more cytoplasmic threads arose which remain connected to the shrunken protoplasts. Often, these threads bore one or more swellings probably due to enclosed bodies. When these shrunken protoplasts were subsequently placed in distilled water, they swelled and burst. In the osmotically stabilized enzyme solution protoplasts could be stable for atleast one week (De Vries and Wessels, 1972).

Isolated protoplasts are physiologically normal, retaining all the properties of intact cells from which they are derived. Comparative studies on protoplasts and whole cells suspended in similar physiological media are few, and in certain instances the data are conflicting. Protein, RNA and DNA synthesis were demonstrated in yeast protoplasts by Hutchinson and Hartwell (1967), with the rates of protein and RNA synthesis similar to those found in the whole cells. DNA synthesis in whole cells was twice the protoplast rate. These findings contrast with earlier experiments where it was found that the synthetic activity of protoplast was considered analogous to the resting cells.

It would be most surprising if the osmotic stabilizer did not have an effect on some, if not all, of the metabolic functions of protoplasts. Amino acids were released from *A. nidulans* protoplasts in both 0.6 M KCl and 0.6 M $MgSO_4$, but a comparative study on mycelium was not made. Kuo and Lampen (1971) carried out a detailed analysis of invertase synthesis and secretion by protoplast and found that both functions were highly sensitive and were inhibited. Several intracellular components leaked out from the protoplasts. Transfer to a medium with low osmolarity reversed the effect and the protoplasts produced and secreted invertase. However, other evidences suggest that both cells and protoplasts of yeast have mechanisms for adaptation to increase in medium osmolarity. Following an initial shrinkage in the medium with higher osmolarity they retain their normal volume. However, metabolic activity remained depressed. Issac (1978) compared the respiratory activities of mycelium and protoplasts of *A. nidulans* over a 3 h period of a lytic digestion. Suspending the mycelium in either 0.6 M KCl or 0.6 M $MgSO_4$ initially enhanced the rate of oxygen uptake and then a gradual decrease followed. The comparative rates were assessed on the basis of protein content. Protoplasts produced with KCl showed an increase in respiratory activity over the mycelium during 3 h, whereas those in $MgSO_4$ had an activity that was much higher at the first hour of lytic digestion, but this declined during the remainder of the experimental period.

1.5 APPLICATIONS OF PROTOPLASTS

a. Protoplasts for Molecular Analysis

Protoplasts were effectively used for the isolation of organelles (nuclei, vacuoles, tonoplasts, ribosomes and mitochondria) from the cells and subsequently for the isolation of mitochondrial DNA (Plate 9) and nuclear DNA.

Protoplasts were the potential source for the isolation of large-scale purified mitochondria and thereby mtDNA. The effective use of protoplasts for the isolation and characterization of mtDNA has been performed in plant pathogens like *V. inaequalis* causing apple scab, *R. solani* causing sheath blight of paddy and *Trichoderma* sp.

b. Protoplasts for Selecting Virulent and a Virulent Cultures of Plant Pathogenic Fungi

Natural selection of virulent and highly pathogenic variants from the regenerated protoplasts culture of *Colletotrichum capsici* was carried out exploiting the variations exhibited among the regenerated colonies. Protoplasts regenerated cultures of *C. capsici* in view of rapid pathogenicity was chosen as the virulent culture (Color Plate 1). Among the tested regenerates C8 & C10 caused complete browning of the fruit proving high virulence. Of the 10

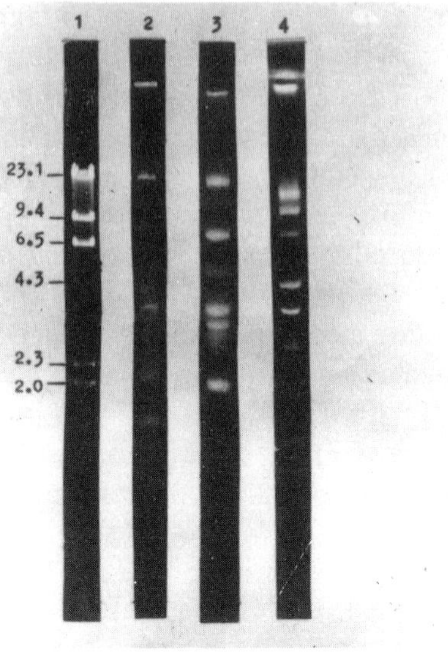

Lane 1 : Marker λ DNA
2 : Sensitive strain
3 : EMS mutant
4 : Adapted mutant

Plate 9. Mitochondrial DNA of *V. inaequals* digested with Hind III restriction enzyme (Mitochondria isolated from protoplasts)

Source: Revathi, 1993, Ph.D. Thesis University of Madras, India, 146.

regenerated protoplasts isolates, 2 of the isolates were very fast growing (C8 and C10) and 2 of the isolates were very slow growing when compared to the parent strain. When the pathogenicity of the protoplast regenerated isolates were checked, all the isolates showed similar results except the two fast growing isolates, which showed rapid colonization and disease incidence, while the two slow growers showed reduced infection and colonization.

This experiment is a proof for the effective utility of protoplasts for screening of the virulent and avirulent strains of plant pathogenic fungi.

c. Protoplasts for Selecting Higher Enzyme Productivity

Protoplast preparation from *Robillarda* sp. strain Y 20, a cellulase producing filamentous fungus, was used for selection of strains with higher enzyme productivity among the protoclones (Kuwabara *et al*., 1989).

Sclerotium rolfsii is known to produce all the three enzymes namely α-amylase, glucoamylose and pullulanase required for hydrolysis of starch. The

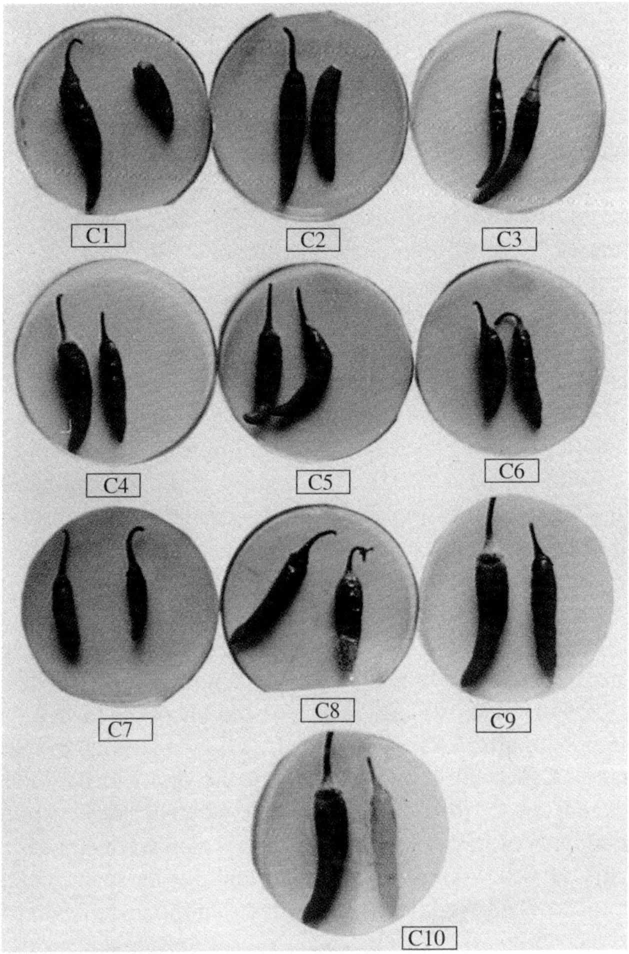

Colour Plate 1

Pathogenicity variations in regenerated protoplast cultures of C. capsici on red chillies.
C1 – C10. Regenerated protoplast cultures

isolated protoplasts are entrapped in calcium alginate gel and the immobilized system was tested for its ability to saccharify various starches and the immobilized preparation retained 57% of its initial activity after three cycles at 30° C at pH 5.0. Though the basidiomycete *Sclerotium rolfsii* (a sclerotial state of *Corticium rolfsii*) has been studied extensively for cellulolytic enzymes, the inability of the organism to produce asexual spores and the mycelial clamp formation during normal growth have hampered its biotechnological applications. Protoplasts of *S. rolfsii* offer a convenient system that can be used to overcome some of these difficulties.

Vacuoles, nuclei, granules, inclusions or particles of various sizes are also found within the protoplast and can sometimes be seen with the phase contrast microscope. Several authors have found that even during lysis of protoplasts the vacuoles can be readily released, and centrifuged down in intact form. Eddy and Williamson (1959) was the first to observe that ultrasonic vibration of yeast protoplasts in certain circumstances released the main vacuoles and smaller vacuoles in a relatively intact form. These vacuoles have been shown to be surrounded by semi-premeable membranes. Shivla *et al.*, (1961) had first demonstrated the utility of *Candida utilis* spheroplasts for obtaining vacuoles after osmotic shock.

Liberation and purification of vacuoles involves floatation from suspensions of lysed protoplasts in the presence of Ficoll (Martile and Wiemken, 1967). The lysis of the protoplasts was achieved decreasing the tonicity of the suspending medium: two volumes of a solution containing 0.025 M Tris-citrate, pH 6.5, 0.025% Triton-x-100 and 1 mM EDTA were added to one volume of a suspension which contained 10^9 protoplasts per ml in 0.7 M buffered mannitol. The lysis was facilitated by a gentle agitation in a Potter-Elvenhjem homogenizer. For the isolation of the liberated vacuoles, Ficoll (Pharmacia Uppsala) was added to the lysed protoplasts to a concentration of 8% (w/v), where the vacuoles assume a density lower than that of the medium. In centrifuge tubes the suspensions were overlayered with two layers, 7.8 and 7.4% of Ficoll. The isolated vacuoles form a white layer on top of the system after 25 min of centrifugation at 2000 g. Preparations of isolated vacuoles showed great variation in the diameters of the spheres, usually between 0.2–2 μm. It is uncertain whether these small spheres represent fragments of large or small vacuoles which were already present in the intact cells. Hydrolytic enzymes are present during specific activities in isolated vacuoles. Therefore it was suggested that the vacuoles represent the lysosome of the yeast cell.

Vacuoles in hyphae and in protoplasts of different species of fungi are easily visible under the phase contrast microscope. The presence of vacuoles in fungal cells and their size are affected by the composition of the growth medium, the age of the culture and possibly growth temperature and other factors. During cell wall digestion, and after the protoplasts have been released, the composition of the suspending medium affects the formation, size and number of the vacuoles seen in the protoplasts. Some protoplasts have several vacuoles of different sizes while others show no apparent vacuole (Garcia Mendoza & Villaneuva, 1967).

Certain conditions required for preparing vacuoles from fungal protoplasts have been established. When a 2 ml volume of *Fusarium culmorum* protoplasts in 0.8 M mannitol was diluted with 5 ml of distilled water, disruption of most of the protoplasts occurred with the release of the cytoplasmic contents. Intact vacuoles could be seen in the suspension and these could be partially recovered by centrifugation. It has also been observed that 0.2 M $MgSO_4$ in 0.1 M phosphate buffer, pH 6.5 allows bursting of protoplasts but maintains the integrity of the vacuoles. The protoplasmic contents are dispersed into the medium and can be eliminated by further washing. On further dilution (between 0.2 M and 0.15 M $MgSO_4$ in the same buffer) the vacuoles swell, ultimately burst and collapse to liberate the contents into the medium. A faint remnant of the tonoplast is left behind. This again adopts a spherical shape, but it is smaller in size than the original vacuole. No attempt has been made to isolate the tonoplasts. Osmotic conditions must be very precise; otherwise the vacuoles were burst. If osmotic changes are not made very gradually, breakage of protoplasts and vacuoles is simultaneous and nothing with a definite organization can be seen.

Bratnicki-Garcia and Lippman (1966) have also described how the dilution of the *Phythophthora* protoplast suspension caused the protoplasts to swell slightly and to burst with disintegration of their membrane. Concomitant with bursting, a spherical vesicle, about the original size of the protoplasts, was frequently formed. Conceivably these vesicles represent swollen cytoplasmic vacuoles and characteristically, protoplast debris remained attached to them from outside.

d. Protoplasts for Developing Laboratory Mutants of *R. solani* Resistant to Fungicides

Stable laboratory mutants resistant to test fungicides were developed using protoplasts. Conveniently, fungal spores were used for producing laboratory mutants by either UV irradiation or EMS mutation. In non-sporulating fungal species, the problem arises during induced mutagenesis. To overcome this difficulty, instead of spores the protoplasts were used for developing mutants. Mutagenesis of protoplasts for fungicide resistance (bavistin) was done by treating with EMS in *R. solani* (Elavarasan, 1996) and the mutagenized protoplasts were plated on PDA supplemented with osmotic stabilizer and bavistin. The regenerating protoplast cultures were grown on a medium with bavistin 5 times the ED_{50} does of the sensitive strain. The stability and resistance of the cultures were checked for ten generations by continuously subculturing on fungicide free medium. Nearly 50% of the regenerating colonies retained their stability. This technique can be applicable to other plant pathogenic and industrially important fungi for developing mutants. Use of protoplasts rather than mycelium or sclerotia or spores greatly simplifies the development of fungicide resistant mutants of *R. solani*. Mutation technique of protoplasts may be broadly used for genetic analysis of fungi. NTG mutagenesis of protoplasts and interspecific fusion in *R. solani* was successfully reported by Hashiba and Yamada (1984).

e. Natural Selection of Potential Antagonistic Strains from the Protoplast Regenerated Cultures of *T. harzianum*

As a biocontrol agent against soil-borne phytopathogenic fungi *T. harzianum* is very promising. The successful use of this species will be greatly enhanced if superior strains could be screened from protoplast regenerated cultures without subjecting it to any artificial mutation or genetic stress. A naturally occurring variant with enhanced antagonistic effect can be selected out from the protoplast regenerated cultures.

Protoplast regenerated isolates of *T. harzianum* showed variation in morphology, growth, sensitivity to potential fungicides and in their antagonistic potential (Table 13). Growth kinetics of regenerated protoplast isolates showed typical variation, though all the protoplast isolates obtained were from a single spore culture of *T. harzianum* (Karpagam, 1994). Protoplast isolates showed a higher rate of growth confirming that the isolated protoplasts are physiologically normal retaining all the properties of the intact cell culture.

Table 13

Tricho-derma harzi-anum	Mycelial Growth	Total protein (mg/g dry wt)	Total DNA (mg/g dn)	Sensitivity to fungicides Bavistin (μM)	Iprodione (mM)	Antagonistic potential Rs	Th	Fo	Th
						Radial growth of mycelia (mm)			
C	Fast	1.1	1.8	18	25	38	90	20	67
P1	Fast	1.1	1.8	40	38	26	90	26	65
P2	Fast	1.3	1.6	21	41	28	90	26	90
P3	Fast	1.3	1.6	22	41	34	90	25	90
P4	Fast	1.2	1.7	23	43	26	90	24	90
P5	Fast	1.2	2.1	14	31	35	90	25	90
P6	Slow	1.4	2.1	40	34	33	90	26	63
P7	Slow	1.1	2.2	33	40	32	90	25	63
P8	Slow	1.1	2.4	21	39	32	90	27	61
P9	Slow	1.3	2.3	18	20	31	90	26	66
P10	Fast	1.4	2.5	40	72	21	90	15	90
P11	Fast	1.5	2.4	19	42	31	90	28	63
P12	Fast	1.1	2.5	35	42	28	90	26	69
P13	Fast	1.1	2.1	27	42	29	90	23	59
P14	Fast	1.0	2.9	20	43	21	90	16	62

P1-P14 regenerated protoplast cultures.
Rs-*R. solani*, Fo.-*Fusarium oxysporum*, Th-*Trichoderma harzianum*

Among the protoplast regenerated isolates screened, one isolate (P 10) of *T. harzianum* showed high antagonistic potential with high tolerance to the fungicides bavistin and iprodione. This isolate was chosen as the potential strain among the protoplast regenerated isolates and subsequently used for the biocontrol of sheath blight of rice caused by *R. solani* and also used as one of the parents for strain improvement through protoplast fusion.

f. For the Preparation of Cell Membranes, Enzyme Activities Associated with Cytoplasmic Membrane

The cytoplasmic membrane of a fungal cell is a distinct and separable structure from the other envelope or cell wall. By gentle enzymatic removal of the fungal cell wall, followed by osmotic rupture of the resulting protoplast in a hypotonic solution, the contents of the cytoplasmic membrane can be removed and the resulting structure, sometimes called 'ghost', cleaned by gentle washing and centrifugations. Studies of such materials showed that fungal (yeast) cell membrane typically consist of protein and lipids with small proportions of carbohydrates and ribonucleic acid. The results so far obtained do not allow the establishment of well defined structural and functional roles of the components identified, but they suggest some unique properties of the fungal plasmalemma.

A study of the activity of the fungal membranes in relation to protein synthesis would be of considerable interest, particularly the extent to which the activity is correlated with the presence of ribosomes since somewhat similar crude membrane fractions from osmotically lysed bacterial protoplasts, which are known to contain ribosomes, are very active in protein synthesis (Cocking, 1984).

Protoplast membranes were usually prepared by lysis of the protoplast using microbial enzyme preparations (Villanueva, 1966).

Few studies have been carried out to study the composition of protoplast membrane from fungi. Boulton (1965) was the first to describe the gross composition of fractions obtained from lysed preparations of *Saccharomyces cereviceae* protoplasts. In an attempt to isolate various cytoplasmic membranes from yeast, protoplasts were lysed under different conditions of temperature, pH, calcium and magnesium ion content. Protoplasts were obtained as described by Eddy and Williamson (1957) using an osmotic stabilizer-mannitol. The protoplasts were kept at 0°C in 10% mannitol solution containing acetate buffer at pH 5. For the preparation of the membrane, protoplasts (40-60 ml as packed volume) up to 48 h old were mixed with ice cold 0.025 M Tris buffer solution (350 ml) at pH 7.2. containing 1 mM $MgCl_2$. The mixture was allowed to stand for 30 min at 0° C and was then centrifuged for 30 min at 20,000 g. The pellet was suspended in a cold solution of the same buffer (50 ml) and gently homogenized by hand in a loose fitting Potter homogenizer. The product was centrifuged for 5 min at

1500 g and the supernatant solution after being separated from the pellet (1.5 p 5 fraction) was centrifuged for 30 min at 20,000 g when a further pellet (20 p 30 fraction) was obtained. The two pellets were next washed by suspending them in the cold buffered solution of $MgCl_2$ (12 ml of each) and centrifuging for 30 min at 20,000 g. The particular fractions so obtained, each containing about 35 mg of protein, were separated and lyophilized.

Boulton (1965) has pointed out that the residues from the above treatment appear to consist principally of membranes forming vesicles of various sizes. The various membraneous structures differ between themselves both in composition and enzyme content. The greater part of the membrane material recovered from the protoplasts was presumably derived neither from the nuclei nor from the mitochondria, but from the outer cytoplasmic membranes and other internal membranes including that of the primary vacuole. It seems likely that internal membranes may contribute about twice as much material as the surface membranes.

Cytoplasmic membranes of yeasts were also prepared by Garcia Mendoza & Villanuneva (1967). Protoplasts of *C. utilis* were obtained by incubating a suspension of cells (5 mg dry wt/ml) at 37°C with shaking in the presence of the 'strepzyme GM' preparation and 0.8 M mannitol as stabilizing agent. The protoplasts were centrifuged at 3000 g for 15 min and washed in the same solution of stabilizer, protoplast membranes were prepared by dilution of the protoplast suspension. An iced solution containing 10^{-4} M Tris buffer (pH 7.2) and 10^{-4} M Mg^{2+} was used and when disruption was complete a solution of the same buffer containing $MgCl_2$ was added to give a final concentration of 10^{-2} M Mg^{2+}. This cation has been shown to prevent further disintegration of the membranes. The membrane fraction was obtained by centrifugation at 15,000 g for 15 min. The dark yellow pellet was purified by washing several times with the same buffer. The cell membrane of *C. utilis* consists largely of protein and lipids each accounting for 40% of the dry weight of the membrane preparations. The total carbohydrate content was found to be 5% glucose, mannose and galactose being the only components. The total nucleic acid content of 1% corresponds to RNA.

Longley *et al.* (1968) made a more detailed analysis of protoplast membranes from another yeast species, *S. cerevisiae*. Growing cells were harvested by centrifugation, washed and suspended in 5 mM citrate-phosphate buffer, pH 5.8, containing 0.8 M mannitol. One volume of snail preparation 40 mg protein/ml was added to 4 vol. of yeast suspension and shaken at 30°C for 3 h. Yeast protoplasts were separated by centrifuging the suspension for 10 min at 0°C at 600 × g. The protoplasts were washed several times with phosphate buffer pH 5.8 containing 0.8 M mannitol and suspended in 100 ml of buffered mannitol (0.8 M) containing 10 mM $MgCl_2$. To prepare protoplast membranes, the protoplasts were resuspended in a small volume of buffered mannitol (0.8 M) containing 10 mM $MgCl_2$. The membranes were washed once more in this buffer and then twice in water. All operations were carried out at 0°C. The membranes were resuspended in 10 ml of water in stoppered tube stored in a vacuum descicator over silica gel at –20°C. These protoplast membranes accounted for 13-20% of the dry weight of the yeast cell. They contained on a weight basis about 39% of lipid, 49% of protein, 6% of sterol and traces of RNA and carbohydrate (glucan × mannan) and whole cells had similar fatty acid composition and contained two major and three minor sterol components.

After sudden and complete lysis of *F. culmorum* protoplasts, a more or less spherical cytoplasmic membrane about the size of the protoplast, showing distinct folds, was sometimes observed (Rodriguez Aguirre *et al.*, 1964). These structures enveloped some cytoplasmic materials in an advanced state of disorganization. Clean and empty membranes were never obtained. Attempts to obtain electron micrographs of these membranes failed owing to difficulties with the usual methods of fixation.

Washed protoplasts free of lytic enzymes were resuspended in cold 0.1 M phosphate buffer pH 6.0. The lysed protoplasts were sedimented at 40,000 g for 15 min in order to collect the membranes. However, instead of a protoplast membrane pellet a large, viscous mass was found. The addition of DNA-ase only partially eliminated the viscous material.

Some other attempts to obtain membrane preparations from filamentous fungi were described by Strunk (1969) using protoplasts of *Polysticus versicolor* prepared by using the gut juice of the snail *Helix pomatia*. Lysis of suspensions of protoplasts followed by centrifugation facilitates the separation of plasmalemma.

g. Enzyme Activities Associated with Cytoplasmic Membranes

An extremely important function of the cell membrane is maintenance of a favourable intracellular environment. The cell membrane, aided by permease system exerts during life a truly remarkable degree of selective permeability upon which the life of the cell depends. It is well known for other systems that many of the enzymic activities of the cell, specially those concerned with the energy-yielding reactions of the cell, undoubtedly take place in the cell membrane.

Early studies on protoplast membranes of 'ghosts' isolated from lysed protoplasts clearly indicated that these structures were biochemically active.

Boulton (1965) has studied the localization of a number of enzymes namely hexokinase, aldolase, pyruvate kinase, phosphoglycerate kinase, triosephosphate dehydrogenase, alcohol dehydrogenase, $NADH_2$ oxidase, $NADH_2$ diaphorase, NADH cytochrome-reductase and ATPase, connected with membrane systems. Some of these enzyme system might arise from contaminating mitochondria. Garcia-Mendoza (personal communication) has also detected ATPase activity associated with a purified membrane preparation.

In purified preparations of plasmalemma isolated from whole cells, various enzymes known to be localized in the ground cytoplasm, in vacuoles or mitochondria are absent (Matile, 1969). The prominent enzyme activity associated with these preparations is that of an Mg^{++} dependent ATPase. A comparatively low specific activity of invertase has been detected occasionally in freshly prepared membranes. The author speculates that the plasma membrane is the place of reactions which result in the conversion of the internal into the external invertase (Gascon *et al.*, 1965).

Chitin synthetase activity of yeast was demonstrated in particles derived from the cell membrane (Kinsky, 1962) and similar results were obtained by Algranati *et al.* (1963) who showed that the cell membrane was the site of synthesis of yeast mannan. Further investigations with these systems would not only give valuable information about the localization of enzymes in the envelopes of mould and yeast, but also contribute to our knowledge of the chemical anatomy of these structures.

h. For the Preparation of Cell-free Extracts

Lysates of protoplasts produced by diluting the suspension medium can be used for investigations of the biochemical potentialities of subcellular particles. Protein synthesis activity of protoplasmic components can be studied, isolation of highly polymerized and biologically active deoxyribonucleic acid in excellent yield may be effected. Enzymes (invertase) can also be isolated from protoplast lysates which sometimes cannot be detected in cell-free extracts prepared by other methods (Gascon *et al.*, 1965).

i. For investigations of Protein and Enzyme Synthesis

Little work has been done on the ability of fungal protoplasts to synthesize constitutive and inducible enzymes but there is no doubt that protoplasts can make proteins and various other compounds as rapidly as intact living cells. De Kloet *et al.* (1961) followed protein synthesis by incubating protoplasts of *S. carlsbergensis*. Incorporation of amino acids was detected and inhibition of this incorporation and of the adaptive enzyme synthesis was observed when respiration was modified. Enzymes such as the ribonuclease apparently produced marked inhibition of protein synthesis when the naked protoplasts were exposed to the nuclease. Van Dam *et al.* (1964) also described a repressor of induced enzyme synthesis in naked protoplasts. The extent of incorporation of carbon form a wide variety of substrates makes it apparent that '*de novo*' synthesis of the cellular constituents does occur in isolated protoplasts of this species.

It has been suggested that induction of some enzymic systems in protoplasts appears to be a rare phenomenon. Volfova *et al.* (1968) using a strain of *Candida lipolytica* which oxidizes hydrocarbon adaptively, showed that protoplasts, unlike the whole cells, are not capable of oxidizing those compounds. These workers showed that their cells cultivated on glucose were able to adapt themselves to hydrocarbon oxidation and that protoplasts were not. The same kind of phenomenon was described by Duercksen (1964) in studying the induction of penicillinase with intact cells and protoplasts of *B. cereus*.

j. For the Study of Osmotic Systems

It is well known that protoplasts are osmotically sensitive and that when they are not encased within the protective, rigid cell wall, they are stable only in media containing high concentrations of impermeant solute. Despite the widespread use of wall-less protoplasts, there still remains an overall vagueness in current knowledge of the processes involved in their osmotic swelling and bursting. Probably the cells most thoroughly studied in regard to osmotic behaviour are mammalian red blood cells. It is difficult to compare the osmotic responses of microbial protoplasts with those of erythrocytes because of our limited knowledge of the former processes.

Intact cells of fungi are assumed to have an osmotic barrier permeable to small molecules, which adhere to the wall and surrounds the protoplasm. This osmotic barrier corresponds to the cytoplasmic membrane. The free protoplasts of fungi are sensitive to osmotic shock and lyse immediately when placed in distilled water. Svihla *et al.* (1961) have determined the optimum conditions for osmotic stability by measuring the release of radioactivity from labelled protoplasts of *C. utilis* placed in several concentrations of KCl. Higher stability of the spheres was found with KCl concentrations of 0.5-1.0 M; Lillehoj and Ottolenghi (1967) have also studied the osmotic properties of *Saccharomyces* protoplasts. This work showed that cells grown in different osmotic pressures behave differently not only in respect to growth (increase in the osmotic pressure caused a decrease in the growth rate) but also in other physiological activities, i.e. respiratory activity of the cells is also affected by the high osmolarity of the medium.

A study of the relative effectiveness of sugars, of short peptides and of amino acids as osmotic stabilizers for bacterial protoplasts indicate that the protoplast membrane can act as a porous differential dialysis membrane and that its effective porosity increased when it is stretched during osmotic swelling. The protoplast membrane also behaves as a highly extensive structure, in contrast to membranes such as those of erythrocytes, and enormous protoplasts could be prepared by slowly dialyzing stabilizing solutes. It appears that when protoplasts swell in hypotonic solutions, their surface membranes may become sufficiently stretched so that they admit stabilizing solutes. There is then a rapid influx of solutes and water resulting in rapid stretching of the membrane and rupture due to a process of brittle fracture. Thus, bursting generally occurs without the membrane becoming fully extended. It appears that bacterial protoplast membrane is physically different from the erythrocyte membrane despite their similar appearances in electron micrographs (Corner and Marquis, 1969). It has been observed that the osmotic fragility of protoplasts varies markedly, depending on the stage of the culture cycle at which cells were harvested. Therefore to obtain consistent results, it is convenient to harvest cells only from cultures in the same stage of growth.

k. For Studies of Permeability

Permeability properties of protoplasts and spheroplasts have been rather widely studied in bacteria. Relatively little work has been done in fungi, although Heredia *et al*. (1968) have exploited the presence of the constitutive hexose transport system in protoplasts of *Saccharomyces cerevisiae* to study its specificity turbidimetrically.

Protoplasts of fungi (mostly yeasts) have proven to be of great value for metabolic studies, mainly those related to sugar transport. Permeability of fungal protoplasts seems to be very similar to those of intact cells. Holter and Ottolenghi (1960) found that when cells of *Saccharomyces* are impermeable to some sugars such as melibiose and sucrose, their protoplasts retain this property. This clearly suggests that the cell membrane acts as a barrier through which some sugars and other compounds are not free to pass. The efficiency of the membranes as a barrier depends presumably on the fungus. In *S. cerevisiae* protoplasts, penetration of some sugars may be catalyzed by a constitutive transport system (Heredia *et al*., 1968). Cirillo (1966) has suggested that the efficiency of such transport is much higher than penetration by simple diffusion. Yeast protoplasts have been extensively used to study the phenomenon of alteration of cellular permeability using antibiotics and other chemicals (Marini *et al*., 1961). Penetration of enzymes (ribonuclease) has also been studied using yeast protoplasts. Whereas ribonuclease does not affect living yeast cells, it can slowly penetrate into protoplasts digesting the volutin in the cytoplasm (Necas, 1971). In contrast Schlenk and Dainko (1966) have described penetration of ribonuclease into intact yeast cells.

Protoplasts of filamentous fungi have been much less used for permeability studies. Elorza *et al*. (1969) using protoplasts of *Aspergillus nidulans*, followed the penetration of a number of compounds and found that at least four transport systems in *A. nidulans*—one for sugar-remain functional after removal of the cell wall. This is consistent with the generalization (Villanueva, 1966) that the permeability properties of fungal protoplasts are similar to those of intact cells.

l. For Growth Studies

Protoplasts have also been used for growth studies. Emerson and Emerson (1958) and Bachman and Bonner (1959) were the first to observe growth of protoplasts obtained from filamentous fungi. However, growth of hyphal protoplasts soon results in regeneration with the ultimate formation of normal mycelium. No growth was ever observed in conidial protoplasts of *F. culmorum* that always resulted in the formation of a germ tube (Villanueva, 1966). Growth of protoplasts has also been observed as a result of fusion of two independent spherical units (Strunk, 1969; Lopez Belmonte *et al.*, 1966). In marked contrast, yeast protoplasts placed in an appropriate medium grow, increasing in DNA and RNA, but do not divide (Tabata *et al.*, 1965; Shockman and Lampen, 1962). Eddy and Williamson (1959) using media inoculated with 10^7 protoplasts per millilitre found that cell nitrogen increased four- to eight-fold in 24 h. The larger increases were obtained when a smaller inoculum was employed.

m. For Test of Resistance to Physical Factors

Protoplasts of fungi being deprived of the protective wall, are susceptible to a variety of physical forces. Sensitivity to a number of factors such as centrifugation. shaking, sonic vibration and ultraviolet irradiation have been tested. Protoplasts can be centrifuged gently and washed, provided they are maintained in appropriate stabilizing solutions. However, the extent of cell damage is proportional to the force employed and is affected by the organism used. Damage to *Fusarium* protoplasts was very small even at 20,000g for 15 min (Rodriguez Aguirre *et al.*, 1964). By contrast centrifugation at 10,000 g for 10 min produced a considerable amount of cell breakage of protoplasts of *C. utilis* (Svihla *et al.*, 1961).

The protoplasts of most fungi appear to be very sensitive to sonic vibration. An exposure of 1-2 min produced total breakage of the osmotic spheres of *Fusarium* and *Canida* species (Rodriguez Aguirre *et al.*, 1964; Svihla *et al.*, 1961).

n. For Studies on Localization of Enzymes

A wide variety of fungal protoplasts have been used from time to time by different workers to study the localization of various enzymes. The role of the cytoplasmic membrane in the secretion of enzymes by micro-organisms has aroused great interest in recent years. Yeast invertase is an interesting system to study this phenomenon. Most of the enzyme is located in the wall. The secretion of invertase by fungal protoplasts has been studied by various workers. Friss and Ottolenghi (1959) showed that protoplasts in species of *Saccharomyces* from sucrose-adapted cells, released more than 74% of their invertase activity. Isolated cell walls of *S. cerevisiae* contain large amounts of invertase but the protoplasts failed to ferment sucrose while still fermenting glucose (Sutton and Lampen, 1962).

The distribution and characteristics of invertase isozymes of yeast and moulds were studies using protoplasts. The results of the distribution of the invertases in *Saccharomyces* and *Neurospora* protoplasts are markedly different. Metzenberg (1964) has shown that in *Neurospora* both isoenzymes can coexist outside the cytoplasmic membrane, although the heavy one predominates inside the protoplasts. Sentandreu *et al.* (1966) confirmed these results working with *C. utilis* protoplasts. In contrast in yeasts the heavy and light invertases correspond to the external and internal enzymes. Only the external invertase contains carbohydrate. The invertase in the protoplast inside the membrane is a smaller form (Gascon and Lampen, 1968).

Related enzymes also appear to be associated with the cell envelope. Complete lack of melibiase was observed in protoplasts of *Saccharomyces* although the enzyme could be detected in intact yeasts (Friss and Ottolenghi, 1959).

Parallel studies to those carried out with invertase enzyme were made with acid phosphates. Experiments with intact and disintegrated cells and application of the protoplast technique have demonstrated that acid phosphatase of baker's yeast is mainly located on the surface of the cell, whereas the alkaline phosphatase is found inside the cell. McLellan and Lampen (1963) found that yeast protoplasts are capable of synthesizing acid phosphatase and the enzyme is released into the medium. Tonino and Steyn Parve (1963) showed that this enzyme is located in the wall of the yeast cells although situated in two different areas of the cell wall. It is thus apparent that yeast wall may accumulate enzymes which have been liberated from the cytoplasm. It has been suggested that the acid phosphatase is of low substrate specificity.

An interesting study was recently made by Nurminen *et al.* (1970) of the enzyme content of the isolated cell walls and of a plasma membrane preparation obtained by centrifugation after enzymic digestion of the cell walls of baker's yeast. The isolated cell walls showed no hexokinase, alkaline phosphatase, esterase of NADH oxidase activity which exist in the interior of the cell. On the other hand considerable amounts of invertase and a variety of phosphatases were found in the isolated cell walls. Enzymic digestion of the cell wall released most of these enzymes but the bulk of the Mg^{2+} dependent adenosine triphosphatase remained in the plasma membrane preparation. This finding suggests that this phosphatase is an enzyme of the cytoplasmic membrane whereas the other enzymes are located in the cell wall outside the plasma membrane. *Suomalainen's* group showed that Mg^{2+} dependent adenosine triphosphatase differs from the phosphatases with pH optima in the range pH 3-4 with regard to location, subtrate specificity and different requirement of activators.

o. For Studies on the Mode of Action of Antibiotics and Surface Active agents

Nystatin and other polyene antibiotics, which are known to bind to the sterols of fungal membranes, produce membrane damage and cell lysis (Lampen *et al.*, 1962; Kinsky, 1963). A number of surface active agents and related substances known to disrupt cytoplasmic membranes were tested against protoplasts of fungi. Digitonin acting on *Aspergillus* and *Fusarium* protoplasts causes immediate lysis (Rodriguez Aguirre, 1965). Sodium dodecylsulphate (0.05%) causes great alternations of the osmotic barrier of protoplasts followed by lysis (Rost and Venner, 1965).

p. For Studies of the Regeneration of the Cell Wall and its Biosynthesis

The protoplasts of fungi appear to have all the synthetic abilities of whole cells from which they are derived, including the ability to make a new cell wall. Fungal protoplasts are very useful for the study of the biosynthesis of the cell wall. As stated by Svoboda *et al.* (1969), it is possible in these protoplasts to trace a gradual construction of single wall components, their arrangement in a complex structure, the regulatory mechanisms of these processes and their relation to the other cell structure.

q. For Studies of Conjugation between Protoplasts

A number of fungi have been used to study the physiology of the conjugation process or cell fusion.

r. For Studies of Spore Formation

When protoplasts of *Saccharomyces carlsbergensis* were inoculated in a medium containing glucose-yeast extract to which 0.6 M KCl was added, formation of spores was observed (Holter and Ottolenghi, 1960). Formation of spores started 14 h after incubation in a starvation medium containing sorbitol and phosphate. About 1-7 spores per cell were observed. The spores burst during the early stages of formation but were resistant when fully formed. Formation of spores was observed in *Schizosaccharomyces pombe* only if the yeast cells were allowed to form zygotes before the cell walls were digested with the lytic enzymes.s

s. DNA transformation in Fungi by Using Protoplasts

Fungal transformation provides an excellent approach to genetic manipulation involving protoplasts. Hinnen *et al.* (1978) have shown transformation in yeast using a chimeric bacterial plasmid called ColE1 with the yeast leu^+ gene incorporated. Uptake was stimulated by the use of PEG. Evidences indicated that the complete plasmid was integrated into the yeast chromosome at several sites. In this experiment the inheritance of the leu^+ phenotype was cytoplasmic.

Transformation of plasmid DNA coding β-tubulin resistance from PSV50 cosmic vector was effectively carried out into the protoplasts of *T. harzianum,* which is highly sensitive to benzimidazole fungicides (Mrinalini, 1997). The effect of PEG on plasmid DNA transformation was also successful. Electroporation method can also be used for plasmid DNA transformation both in the presence and absence of PEG.

CHAPTER 2

Regeneration and Reversion of Protoplasts

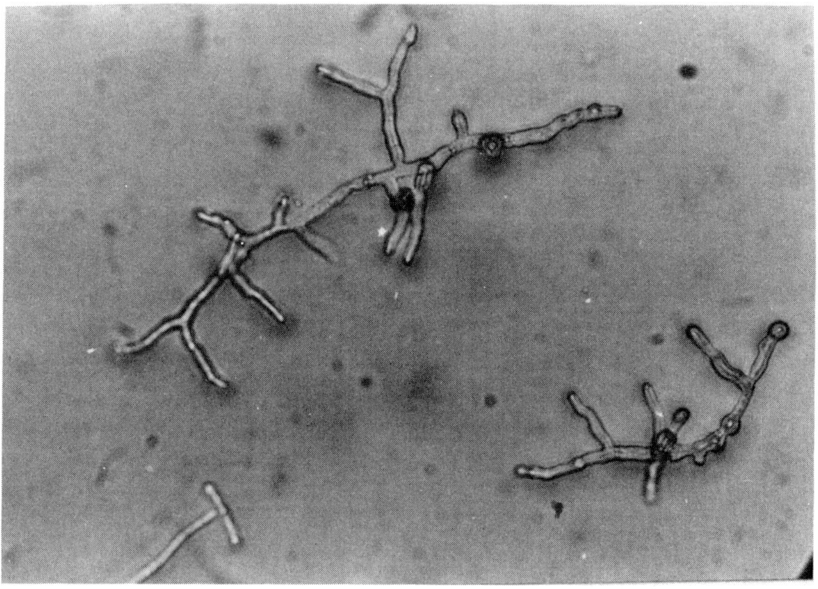

Each viable protoplast is considered an individual living entity containing nucleus which restores the genetic information of the parent organism. Cell wall biogenesis is coded by chromosomal genes, hence only nucleated protoplasts can successfully synthesize the cell wall and subsequently regenerate and revert to parent culture. The ability of cell wall rebuilding and reversion to parental form is termed regeneration and reversion of protoplasts.

2.1 REGENERATION AND REVERSION

The fungal protoplast rapidly regenerates a wall (regeneration) and reverts to the original mode of growth (reversion). Most studies on wall regeneration have been carried out with *Saccharomyces cerevisiae*. In the case of yeast, there are 2 major draw backs.

1. It is very difficult to obtain protoplasts without any wall remnants (Bacon *et al.,* 1969; Darling *et al.,* 1969).
2. The protoplasts failed to regenerate a wall of normal composition and structure in liquid medium (Kreger & Kopecka 1975) nor do they revert to normal budding cells Necas, 1971).

The initial sign of growth of regenerated protoplasts as bud-like protrusions is very common in fungi. It is plausible that the budding might be due to an imbalance in the ratio between cell volume and cell wall materials during the initial stage of regeneration. Studies on regeneration and reversion of protoplasts in filamentous fungi are interesting because these systems provide an opportunity to observe the emergence of hyphal structure with their highly organized apical growth centres. These hyphae always arise from more or less spherical cells after regeneration of walls around the protoplast. The cell wall materials were deposited outside the plasmalemma starting with a very coarse structure. Loose, thick, aberrant cell wall appeared and it became gradually denser with increasing regeneration time. The thickness of regenerated cell wall was approximately 500 nm and sometimes as thick as 1300 nm. The thickness of the cell wall of normal hyphae after 16 h cultivation was approximately 100 nm. The wall of regenerating protoplasts was rich in electron dense material compared with the wall of normal hyphae. The thick regenerated cell wall is composed of several layers.

The first and foremost important process in regeneration of protoplast is the synthesis of the rigid cell wall. Most studies on the regeneration of protoplast concerned simple microscopical observations (Villanueva and Acha, 1971; Peberdy and Gibson, 1971; Berliner *et al.,* 1972; Dooijewaard-Kloosterziel *et al.,* 1973; Peberdy and Buckley, 1973; Siestsma and DeBoer, 1973; Anne *et al.,* 1974; Annamalai and Lalithakumari, 1991; Revathi and Lalithakumari, 1992; Mrinalini and Lalithakumari, 1993; Karpagam, 1994; Vijayapalani, 1995; Elavarasan, 1996).

In general, it appears from the above studies that after the removal of lytic enzyme used in protoplast formation, a carbon source is required to obtain regeneration or the deposition of some sort of a wall around the protoplast. Subsequent development is then characterized by the generation of abnormally shaped structure, some of which may eventually develop into a normal looking hypha. Use of solid medium for regeneration such as agar, seems to increase the incidence of hypha emerging directly from the regenerated protoplasts.

The reason for this is obscure. In the case of *Schizophyllum commune* it has been reported that up to 50% of regenerating protoplasts directly produced hyphae. Peberdy (1976) showed the heterogeneity of protoplasts derived from filamentous fungi with respect to their nuclei and other organelles. These differences are reflected in their regeneration and reversion pattern. De Vries and Wessels (1972 and 1975) showed that centrifugation with $MgSO_4$ as osmotic stabilizer resulted in 2 classes or protoplasts, namely non-vacuolated protoplast without nuclei and vacuolated protoplast with nuclei, which influenced regeneration.

The non-vacuolated protoplasts showed little regeneration capacity while vacuolated protoplast showed more than 50% regeneration of cell wall with reversion to hyphal mode of growth. The highly vacuolated protoplast on staining with Giemsa, revealed 70% of protoplasts containing one or more nuclei. This would suggest that although the presence of a nucleus is a prerequisite for the occurrence of reversion to normal hyphae, all the nucleated protoplasts do not show reversion. Peberdy and Buckley (1973) in their studies on applying optical brightner observed 2 phases in the capacity of absorbance of tinopol during regeneration of *Aspergillus nidulans* protoplasts. There was an initial phase of low adsorption followed by a second phase with a high capacity of adsorption of the brightner seems to coincide with the formation and growth of hyphae from the majority of regenerating structures. This increased capacity of adsorption of flourescent brightner was correlated to an increased rate of β-glucan synthesis (De Vries and Wessels, 1975).

When protoplasts are suspended in an osmotically stabilized nutrient medium, part of the population starts to synthesize a new cell wall and eventually return to normal hyphal form. This is termed as regeneration. These protoplasts regenerate into the hyphal form. The process of germination or regeneration of protoplasts is an important factor. A careful analysis of the regeneration of the protoplasts had given interesting observations. The protoplasts on incubation with a nutrient medium supplemented with respective osmotic stabilizer put forth germtube-like structure similar to spore germination. The extension or elongation of this primary germ tube or hyphae further depends on the physiology of the protoplasts. The mode of hyphal regeneration varies in filamentous fungi. Three basic patterns of regeneration were observed (Fig. 13).

In the first type, protoplasts produced yeast-like buds and developed into irregularly shaped chains of cells (Plate 10a and Fig. 13a). Cytoplasmic contents became lucent under phase-contrast microscopy and finally cells autolyzed without hyphal development (incomplete regeneration). Autolysis of germtube is common at different stages of germtube elongation. The autolysis could occur even after the withdrawal of osmotic stabilizer from the growth medium. The second type of regeneration was similar to the first, but after development of an irregular chain of cells, a germtube-like hypha protruded from the cell distal to the original protoplast and further a normal apical type of hyphal growth was restored (Plate 10b). A very interesting mode of regeneration was observed in the third type. The regenerating

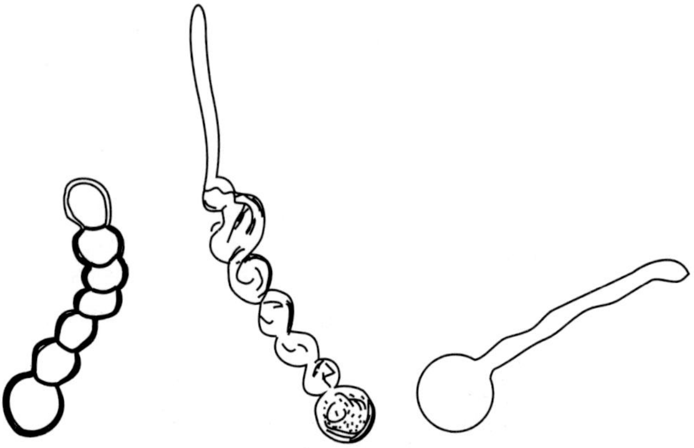

Fig. 13. Schematic representation of hyphal regeneration from protoplasts of filamentous fungi. (a) 10 h regeneration (incomplete regeneration). Chain of irregularly shaped cells and finally autolysed without formation of hypha.
(b) 20 h regeneration. Protoplast formed in the chain of irregularly shaped cells and finally protruded germ tube like hypha from the cell distal to the original protoplast.
(c) 16 h regeneration. Protoplast remain spherical without forming buds. A germ tube like hypha was formed directly from the single spherical cell.
Source: Tanaka et al. 1981.

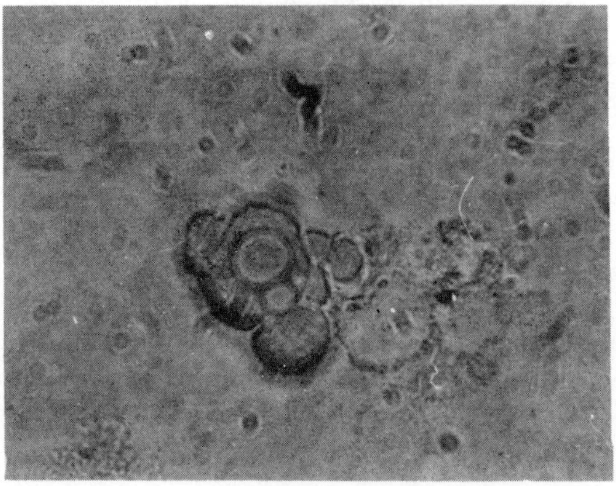

Plate 10a. *V. inaequalis* protoplasts produced least-like buds which subsequently autolyzed.
Source: Vijayapalani, 1995, Ph.D. Thesis, University of Madras, India, 212.

protoplasts remained spherical for a long time (usually more than 10 h) and they could not be distinguished from non-regenerating protoplasts under a phase-contrast microscope. A germtube like hypha protruded directly from the spherical cell (Plate 10c) and developed into normal hyphal structure (Plate 10b, Fig. 13b).

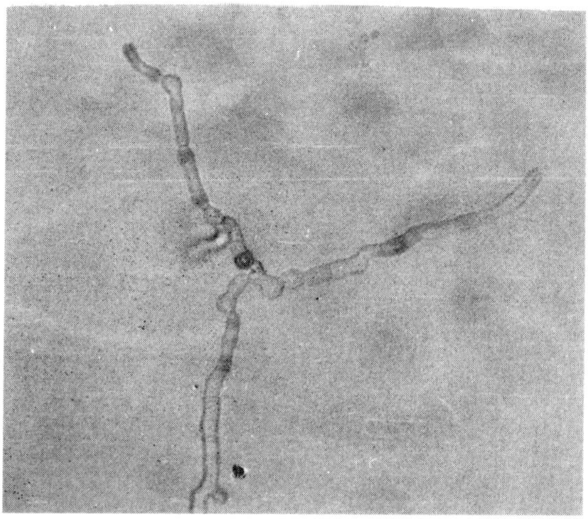

Plate 10b. Irregularly shaped chain of cells or yeast-like buds of *V. inaequalis* giving rise to normal hypha.

Source: Vijayapalani, 1995. Ph.D. Thesis, University of Madras, India, 221.

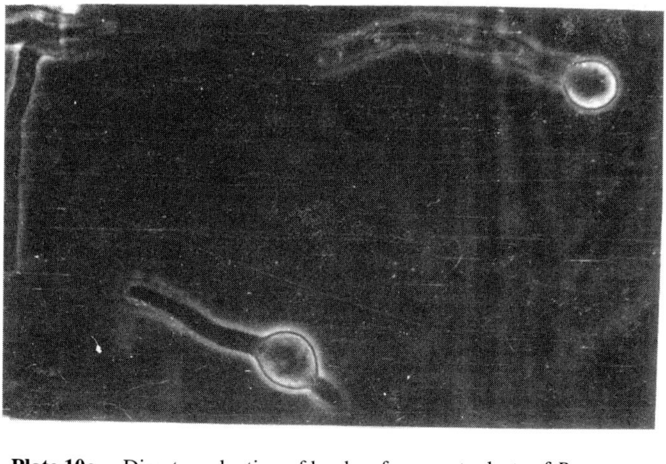

Plate 10c. Direct production of hyphae from protoplasts of *B. oryzae*.

Source: Annamalai and Lalithakumari, 1991. Journal Plant disease and Protection '98: 197-204.

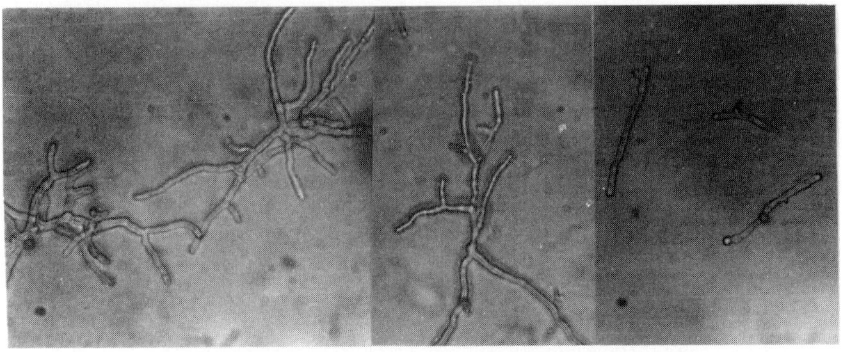

Regeneration forming hyphae directly from the protoplasts.

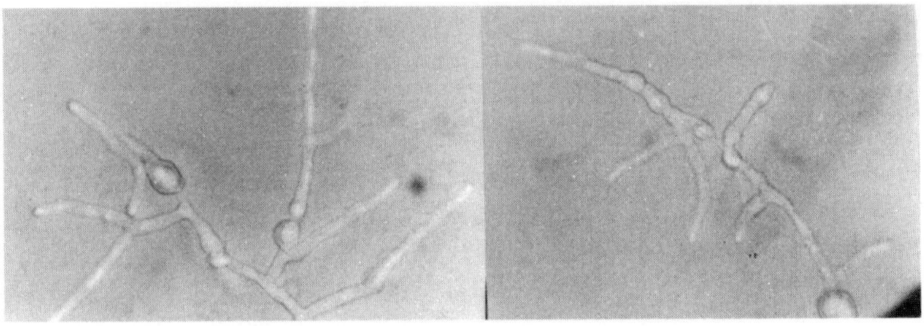

From chains of buds normal hyphae emerge.

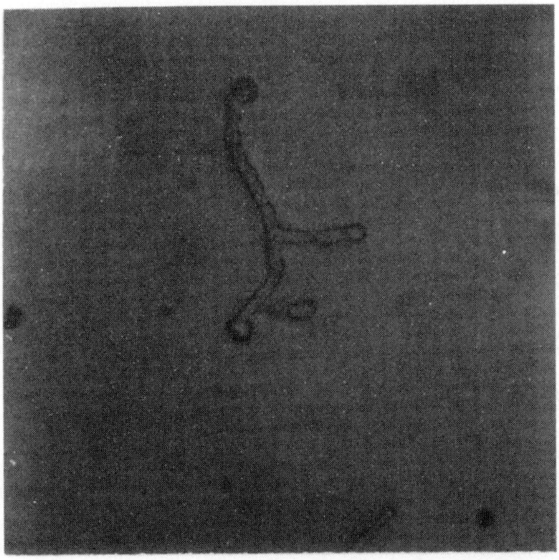

Chains of irregularly shaped cells without formation of hyphae.

Plate 10d. Phase contrast microscopic observations hyphal regeneration from protoplasts of *Trichothecium roseum*.

Source: D. Lalitha Kumari, 1999 (Unpublished)

The regenerating protoplast of type 3 (Plate 10c + Fig. 13c) is especially interesting. Nuclear division and synthesis of cytoplasmic contents were somehow suppressed and only the cell wall was actively synthesized. The synthesis of cell wall probably continued until the protoplasts established conditions for apical growth, either in the cell wall or in the cytoplasm near the plasmalemma. Protrusion of a germtube-like hypha must have occurred after acquiring the apical growth condition. Plate 10d shows three different types of regeneration of protoplasts of *Trichotherium roseum.* Yeast protoplasts regenerate by bidding (Plate 10e).

2.2 FACTORS AFFECTING REGENERATION OF PROTOPLASTS

Important parameters in the regeneration of protoplasts are viability, the capacity to synthesize cell walls and the retention of properties of the parent cell. Factors influencing regeneration of protoplasts are the nature of osmotic stabilizer, the concentration of osmotic stabilizer, pH, temperature and the physical state of regeneration medium. Published data relate mostly to reversion and majority by viable count determinations following plating of protoplasts in osmotically stabilized agar media. A reversion frequency of 100% has never been reported by using this approach and the frequency can be quite variable for any single species. Studies with *Fusarium culmorum* (Lopez-Belmonte, *et al.,* 1966) showed that reversion frequency was influenced by the carbon source in regeneration medium. In *A. nidulans,* Issac (1978) had reported that, reversion frequencies were in the range of 10-30%

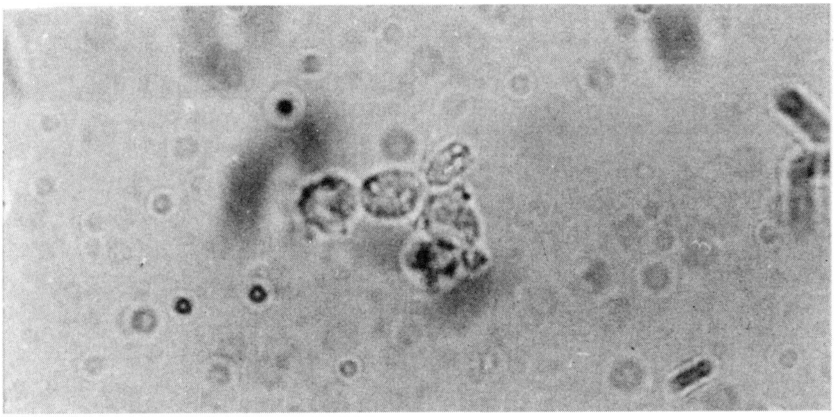

Plate 10e. Yeast protoplast regeneration.
Source: D. Lalithakumari, 1999 (Unpublished)

and no difference was found with defined and complex media. Optimum conditions for protoplast regeneration of *T. longibrachiatum* were glucose-mineral regeneration medium with 0.5 M KCl and pH 6.0 at 30°C (Kumari and Panda, 1993). Osmotic stabilizers like glucose, NaCl and KCl in regeneration medium were found to induce high regeneration percentage of protoplasts 65, 65-75 and 70-75% respectively, in *Talaromyces flavus* (Santos and De Melo 1991). Largest amount of regenerating protoplasts was recorded on using 0.7 M mannitol as stabilizer in *T. harzianum* (Tashpulatov *et al.,* 1991). Revathi and Lalithakumari (1992) have reported that PDYEA amended with 0.6 M sorbitol-sucrose was effective for regeneration of protoplasts of *V. inaequalis*. Protoplasts of *P. oryzae* amended in CZYMAM with 0.6 M KCl exhibited rapid growth (Kumari and Lalithakumari 1987) in CZYAM with 0.6 M sucrose-mannitol is best for *B. oryzae* protoplasts (Annamalai and Lalithakumari, 1991) and PDA with 0.6 M KCl is good for *T. harzianum* (Mrinalini, 1997) protoplasts regeneration.

a. Physical State of Media on Regeneration

Regeneration of protoplasts in liquid medium is very poor. Necas (1971) studied the regeneration of *Saccharomyces cereviseae* protoplasts in liquid medium and reported that protoplasts did not regenerate in liquid medium. But on the other hand 100% regeneration was achieved in 24 h by cultivation in the same medium solidified by addition of 15% gelatin. Any other substance used to solidify the medium did not give satisfactory results. Garcia *et al.* (1966) have reported successful regeneration of *Fusarium culmorum* protoplasts in semisolid agar (0.2% w/v) but reversion was also observed on incubation of protoplasts in solid (2% agar) gelatin (0.2 to 20%). Solid media allowed regeneration of *F. culmorum* protoplasts even on the surface. Under the optimal conditions for regeneration (2% sucrose, 10% sorbose) not more than 3% regeneration was obtained using gelatin (0.2-20%). Only plate surfaces solidified with 10% gelatin and seeded with protoplasts showed better rate of regeneration. On the other hand, use of 0.2% agar instead of gelatin gave over 80% regeneration. The reversion to mycelial forms of *F. culmorum* from spherical protoplasts was followed conveniently in liquid medium, using microdroplets of regeneration medium containing sucrose + sorbose. In this medium, the uniformity of external pressure, the elasticity of the newly formed membrance and interferencial tension, all tended to conserve the more or less spherical shape of the protoplast. The development of regenerated masses from single protoplast during the initial stages were similar to the regeneration in semisolid medium. A number of protruberances and growing spheres were also formed. At times it was seen that local points of weakness where the growing protoplast pinched off a number of small blebs, which sometimes gave amoeboid forms. During the very first stages of regeneration the regenerated forms were osmotically fragile, but once the amoeboid and elongated forms were formed the osmotic sensitivity partially disappeared.

Regeneration of *B. oryzae* protoplasts were performed on 4 different solid media viz. water agar with osmotic stabilizer, water agar without osmotic stabilizer, Czapek's (Dox) yeast extract agar and potato dextrose agar (PDA) (Annamalai, 1989). Either 0.6 M mannitol and 0.6 M sucrose or 0.6 M sucrose-mannitol was used as osmotic stabilizer. For all combinations of the regeneration media, between 50-100 protoplasts were placed on sterile cellophane over solid medium on each of the five agar plates and incubated at 25°C for 3 days (Table 14).

Table 14 Regeneration of protoplasts of *B. oryzae* on different solid media

Medium	Per cent of regenerated protoplasts after		
	24 h	48 h	72 h
Water agar (2%)	–	–	–
Water agar (2%) + 0.6 M Mannitol	5	8	8
CZYAM + Sucrose (0.6 M) & Mannitol (0.6 M)	25	45	47
PDA + Sucrose (0.6 M) & Mannitol (0.6 M)	20	42	45

Source: *Annamalai, 1989.* Ph.D Thesis, Univ. of Madras, India.

The protoplasts had the potential to regenerate on media amended with osmotic stablizers. CZYAM and PDA amended with 0.6 M sucrose-mannitol recorded higher per cent of regenerated protoplasts than the other media amended with osmotic stabilizers. After 24 h, 25 per cent of the protoplasts regenerated on CZYAM media containing 0.6 M sucrose-mannitol. At 48 h, the regenerated colonies increased to 45 per cent but further incubation did not result in the increase of regenerated colonies. The protoplast sincubated on water agar with 0.6 M mannitol, showed only 8% of regeneration and produced viable colonies after 48 and 72 h. Protoplasts did not regenerate on media without osmotic stabilizers. Regenerating protoplasts directly produced germtubes in all media containing osmotic stabilizers. The importance of solid medium over liquid medium (Table 15) for successful regeneration of protoplasts of *V. inaequalis* was reported (Vijayapalani, 1995).

Though both liquid and solid medium supplemented with 0.6 M sucrose + 0.6 M sorbitol favoured protoplasts regeneration, solid medium with 2% agar recorded maximum regeneration.

Table 15 Effect of physical state of the medium on the regeneration of protoplast of *V. inaequalis*

Medium 0.6 M sucrose + 0.6 M sorbitol	Regeneration frequency (%) (72 h incubation)
Liquid Solid (agar %)	12.7
1.0	0
1.5	0
2.0	33
2.5	31
3.0	0

Source: Vijayapalani, 1995. Ph.D. Thesis, Univ. of Madras, pg 97-225

Mrinalini (1997) showed variation in regeneration frequency of protoplasts of *T. harzianum* (Th) and *T. longibrachiatum* (Tl) on different regeneration media (Table 16a and b).

Table 16a Regeneration of protoplasts of the two species of *Trichoderma* on water agar medium

Strain	Different osmotica regeneration (%)			1.4 M $MgSO_4$ & 50 mM Sodium citrate
	0.6 M KCl	0.6 M NH_4Cl	0.6 M Sucrose & 0.6 M Sorbitol	
Th (18 h)	6	2	1	0
Th (36 h)	9	3	2	1
Tl (18 h)	3	2	2	0
Tl (36 h)	5	3	3	2

Source: Mrinalini, 1997. Ph.D. Thesis, Univ. of Madras, pg: 66-102

Table 16b Regeneration of protoplasts of the two species of *Trichoderma* on potato dextrose agar

Strain	Different osmotica regeneration (%)			1.4 M $MgSO_4$ & 50 mM Sodium citrate
	0.6 M KCl	0.6 M NH_4Cl	0.6 M Sucrose & 0.6 M Sorbitol	
Th (18 h)	29	11	14	7
Th (36 h)	53	27	23	20
Tl (18 h)	33	19	11	11
Tl (36 h)	48	25	23	24

Source: Mrinalini, 1997. Ph.D Thesis, Univ. of Madras, pg-66-102

Regeneration was not achieved in liquid media. The compact support provided by solid media was lacking when liquid media were used and so regeneration and reversion of protoplasts were incomplete and unstable in liquid medium. Regeneration frequency also varied with different osmotica in the medium. PDA medium was better over water agar medium.

Agar concentration and pH of medium also play a significant role on the hyphal, budding, budding and hyphal type of regeneration in *P. oryzae* (Asai *et al.,* 1986).

Quantitative thin layer-agar plating

Basal growth medium (BGM) contained 3% yeast extract, 1% peptone, 2% heart infusion broth and 4% glucose; the latter was autoclaved separately. As medium for stabilizing protoplasts, BGM containing 3% NaCl (stabilizing BGM) was employed. Protoplasts were quite stable in stabilizing BGM. The modified and stabilizing Czapek-Dox's medium (pH 6.8) tested as synthetic medium had the following composition (g/L); KCl 0.5; $MgSO_4$ 0.5; $FeSO_4$ 0.01; K_2HPO_4 1; $NaNO_3$ 3; glucose 40 and NaCl 45. For serial dilution of protoplasts, stabilizing CPBS was employed. The concentration of purified protoplast suspension was determined by measurement of the number of protoplasts with a Burker-Turk tube haemocytometer under a phase-contrast microscope. To study regeneration of protoplasts in solid medium, a sterile 2% solution of special agar (from Difco) dissolved in stabilizing CPBA (stabilizing agar) was prepared and kept in a water bath at 47°C before use.

ii. Effect of modifications in carbon and nitrogen sources of the growth medium on regeneration

Since *F. culmorum* grows very well in Czapek medium, this was used as basal medium for the regeneration of protoplasts from mycelium. The best results were obtained with media containing the salts of the Czapek medium + 10% sorbose, 2% sucrose, 0.2% agar. Concentrations of sorbose between 2 and 5% produced little regeneration and some lysis of protoplasts. Concentration of 10-12.5% were optimal and higher concentrations gave somewhat lower degrees of regeneration. The sole difference found was the increasing refringency of the protoplast contents as the sorbose concentration was increased. No other sugar tested could replace sorbose, but sorbitol replaced sorbose. Neither glucose nor fructose could replace the sucrose which was used in admixture with sorbose. But glucose + fructose satisfactorily replaced sucrose. Increasing the concentrations of surcorse and decreasing those of sorbose gave somewhat lower degrees of regeneration. Impurities in the sorbose preparation also affect protoplast regeneration. The influence of the nitrogen source of regeneration media was examined by replacing the sodium nitrate of Czapek medium by different nitrogen compounds (NH_4NO_3, $(NH_4)_2SO_4$, glutamate, glycine, asparagine, casein and peptone). Casein was the best nitrogen source, but there are no great differences among the other organic and inorganic nitrogen sources. The regeneration frequency of 1.5% of *V. diplasia* was observed in the presence of glucose and peptone in the complete medium. Substitution of peptone with asparagine increased reversion frequency confirming that the presence of carbon and nitrogen source is essential for protoplast regeneration (Khanna *et al.,* 1991). Santiago (1983) achieved about 4% reversion of protoplasts of *Volvariella* on a medium supplemented with glucose and asparagine.

iii. Protoplast concentration on regeneration

In quantitative studies on regeneration of protoplasts, the correlation between the number of colonies and the protoplast concentration is very important.

In some cases, a straight line relationship exists between the dilution of protoplast suspension inoculated and the number of colonies observed. This indicates that individual colonies were produced by regeneration of a single protoplast.

iv. Effect of temperature, aeration and pH on regeneration

The influence of several environmental and chemical factors on regeneration from protoplasts has been reported by Garcia *et al.,* (1966). Under the optimal conditions temperatures between 18 and 33°C did not significantly affect the number of protoplasts regeneration in semi-solid or liquid medium. When incubation was at 28°C in liquid media aeration had no noticeable effect and shaken and static cultures showed about the same degree of regeneration of *F. culmorum*. Effect of temperature on the release of protoplasts of *V. inaequalis* (Table 17) was reported also by Vijayapalani (1995).

Different pH of the medium show direct effect on regeneration of protoplasts. When protoplasts of *F. culmorum* were incubated at 28°C statically the maximum regeneration (70%) was at about pH 6.7. At pH 5, 50% of the protoplasts regenerated; at pH 8 only 12% of the protoplasts reverted to normal mycellium. Above pH 8 regeneration did not occur. At pH 3, 25% of the protoplasts regenerated. The curve relating regeneration and pH value was linear from pH 3 to 6.5 and decreased sharply above pH 7 confirming the strong influence of pH on regeneration of protoplasts.

Table 17 Effect of temperature on the release of protoplasts of *Venturia inaequalis*

Incubation condition	No. of protoplasts (× 10^3) / mg f.wt. of mycelium
Temperature (°C)	
30	1.1
32	1.4
34	1.5
36	3.1
38	1.7
40	1.0

Source: Vijayapalani, 1995, Ph.D Thesis University of Madras 97-225.

b. Other Factors affecting Regeneration of Protoplasts

Regeneration of protoplasts could not be demonstrated in broth media such as stabilizing BGM or medium containing 20% calf serum and in these media, protoplasts lysed within few hours. After 24 h, a few protoplasts showed irregular growth forms, similar to those reported by Gascon *et al.* (1965) for various yeast and mold protoplasts. However, most abnormal cells lysed within an incubation period of 24 h. Protoplasts embedded in thin layers of stabilizing agar could regenerate in stabilizing BGM of pH 5.5 to 8 at 20-30°C.

i. Amino acid pool

Importance of amino acid in the regeneration of protoplasts has been well documented in ergotamine production in *Claviceps purpurea* (Basett *et al.*, 1973; Floss, 1976). The analysis of the amino acid pool of the protoplasts of *C. purpurea* in ergotamine production showed differences in the rate of uptake of amino acids. The analysis of the amino acid pool of the protoplasts of *C. purpurea* showed that, alanine, the precursor of ergot peptides was present in highest concentration followed by proline and phenylalanine. The low incorporation of amino acid was due to a partial inability to penetrate the cell. The rate of uptake in the control mycelium was higher (92%) in 90 min, while it was only 42% by the protoplasts.

ii. Biotin and Lipoic acid

Protoplasts are able to develop moderately well into colonies after the addition of biotin and lipoic acid to the modified Czapek-Dox's medium. However, regeneration of protoplasts could not be demonstrated.

iii. Influence of lytic enzymes on protoplasts regeneration

Lytic enzymes used for the release of protoplasts play a very important role on regeneration. Regeneration of the protoplasts of *F. culmorum* obtained by means of the snail enzyme preparation follows steps very similar, nearly identical, to those described for protoplasts obtained with strepzyme RA. The only significant difference was the formation of a large number of convoluted forms in the protoplasts obtained by strepzyme RA. Regeneration studies with protoplasts obtained with snail enzyme preparations were often complicated by bacterial contamination. Attempts made to obtain protoplasts of *F. culmorum* by growing the organism on a medium containing high concentrations of sorbose, according to Hamilton and Calvet (1964), were unsuccessful.

c. Ultrastructural Observations on Regeneration and Reversion

Extensive studies on wall ultrastructure and composition have been made with only a few organisms, with yeasts again claiming attention. The incomplete wall formed by *Saccharomyces* sp. protoplasts cultured in liquid medium examined in several laboratories revealed that the wall takes the form of a network of microfibrils. Formation of this network proceeds for 6 h and is then repressed. It is this abnormality of the wall, with the absence of an amorphous component that is assumed to be the cause of inability of the protoplast to undergo reversion. The amorphous component or its precursors, or the enzymes involved in its synthesis, may be lost into the medium through the microfibrillar net, whereas they can be retained in the net in protoplasts in a solid medium. Anucleated protoplasts of *S. cerevisiae,* derived from buds, do not produce the fibrillar network confirming the importance of nucleus in wall synthesis.

i. Lomasomes

Lomasomes (border bodies) were reported by Girbardt (1969) and named by Moore and McAlear (1961). Since then lomasomes and similar structures have been reported in many fungi. Tanaka *et al.,* (1981) observed many vesicular structures (lomasomes) in regenerated cell walls of *P. oryzae.* Structures termed lomasomes by various investigators differ from each other morphologically (Bracker, 1971). These structures were also reported by Heath and Greenwood (1970) in the cyst wall of *Dictyuchus sterile.*

A variety of functions have been attributed to lomasomes, and their role in cell wall formation is suggested by several investigators (Bracker, 1971).

ii. Golgi apparatus

Typical Golgi apparatus has been reported only in limited number of fungal species. In *P. oryzae,* typical Golgi apparatus has not been reported. However, structures resembling Golgi apparatus, i.e., stacks of cisternae appeared in the regenerating protoplasts. The structures were not observed after completion of regeneration after a long period (Tanaka *et al.* 1981). Golgi apparatus-like structures were frequently observed in the protoplasts of *P. oryzae* at the beginning of regeneration (2-4 h). However, this organelle was not found in the protoplast after 8 h of regeneration or longer. Havelkova (1969) also reported that extensive formation of dictyosomes occurred at the start of regeneration of *Schizosaccharomyces pombe* protoplasts and that the number of dictyosomes decreased as synthesis of the new cell wall progressed. If this organelle function to synthesize cell wall materials, the precursors are synthesized in the early stage of regeneration should be stored in the cytoplasm. The amorphous structures could be the organelle, which stores the cell wall precursors and which may be analogous to the apical wall vesicles reported in various species of filamentous fungi (Grove, 1978). The appearance of electron transparent granules of *Cunninghamella elegans* was composed of non-sulfated, slightly acidic polysaccharide (Hawker *et al.*, 1970). A morphologically similar structure was reported in the conidal cytoplasm of *Scopulariopsis brevicalis* (Cole and Aldrich, 1971; Hammil, 1972). Typical Golgi apparatus (dictyosomes) has been reported in a limited number of fungal species. During regeneration, numerous electron-transparent amorphous structures appeared in the cytoplasm and often near the plasmalemma. Some of them resembled vacuoles but they were not bounded by tonoplast. Sometimes they appeared as aggregates of granules or vesicles of about 80 to 200 nm in diameter. The structure appeared to increase as the Golgi-like structures decreased in number and disappeared.

iii. Mitochondria

In the cytoplasm of most protoplasts of *P. chrysogenum* numerous mitochondria were observed. Mitochondria were ovoid and contained a number of long filiform cristae.

iv. Ribosomes

In protoplasts of *P. chrysogenum* ribosomes were abundant though some were deprived of them. Ribosomes distribution reflected the origin of the protoplasts. Ribosomes rich protoplasts were assumed to originate from young hyphal tips with high biosynthetic activity (Trinci, 1978). Cytoplasm with dispersed ribosomes arises from older and less active region.

d. Ultrastructure and Localization of Cell Wall Polymers during Regeneration

During wall regeneration the cytoplasmic density was often much higher than prior to regeneration. This appeared to be mainly due to accumulation of ribosomes, mitochondria and rough endoplasmic reticulum. Cytoplasmic vesicles were sparse in regenerating protoplasts as in isolated protoplasts. Noteworthy is the scarcity of vesicles in regenerating protoplasts of S. *commune*. They only are abundant at the site of bud and hyphal tube cemergence and in the apex of extending hypae. The possible role of these vesicles in hyphal extension has been extensively discussed (Girbardt, 1969; Grove and Bracker, 1970; Bartnicki-Garcia, 1973) but remains unsolved. There is evidence that the vesicles fuse with the plasmalemma. In this way they would contribute to the extension of plasmalemma and the contents discharged could somehow contribute to wall synthesis. Vacuoles mostly were irregularly shaped and appeared to decrease in size during regeneration.

Glycogen masses which stained heavily with the PA-TCH-SP reagent were abundant in the cytoplasm of most cells and often had a peripheral position. Notable was the association of glycogen with lipid granules. Prior to regeneration, the plasmalemma of the regenerating protoplasts showed a strong affinity to the Thiery reagent.

Observations at high magnification sometimes showed a bilateral staining of the plasmalemma. The plasmalemma of many regenerating protoplasts exhibited invaginations, which varied in complexity and also showed high affinity to Thiery stain. Other cellular membranes showed low affinity to the stain.

After 3 h of regeneration, in *S. commune* the walls of most cells were loosely organized. These walls often differed markedly in thickness and appeared to consist mainly of loose aggregates of fibres with irregular outlines. Such fibers were also found in the surrounding medium suggesting that the fibers easily detached from the cell surface, especially during wall isolation. After 9 h of regeneration, the walls of most primary cells consisted of two layers, i.e. a fluffy, loosely organized outer layer as referred to above and a densely packed layer at the inside. There was no sharp boundary between the two layers. The wall showed a low affinity to staining with uranyl acetate/lead citrate. However, prolonged staining revealed thin fibers in both wall layers. Staining with Thiery reagent differentiated between the two wall layers; the outer fluffy layer remained unstained whereas the inner layer stained heavily. This inner stained layer was very thin or absent after 3 h of culture but was well developed in most cells after 9 h. After 9 h of regeneration only a small proportion of the cells lacked PA-TCH-SP stained inner wall layer and were surrounded only by a fluffy coat of unstainable material.

i. Emergence of buds and hyphae

After 9 h of regeneration many cells were already engaged in the formation of bud-like structures (buds) or hyphal tubes. Buds resembled primary cells in having a fluffy outer wall layer. Hypahe of reverted protoplasts lacked the fluffy coat and instead had a compact wall. Buds and hyphae were only initiated in primary cells possessing the PA-TCH-SP stained inner wall layer. The PA-TCH-SP stained portion of the wall of most emerging buds and hyphae had an irregular outline, rather thin and continuous with only the inner portion of the stained inner wall layer of the primary cell.

Contrary to the sparse and irregular distribution of cytoplasmic vesicles in primary cells, vesicles (about 100 nm diameter) were commonly found in the tips of emerged hyphae and in the initials of either hyphal apices or buds. Even the very inception of these initials could be traced by the presence of such vesicles. The origin of these vesicles is not clear.

e Examination of Freeze-Etched and Negatively Stained Preparations of Regenerating Protoplasts

Prior to regeneration, the outer surface of the plasmalemma of the protoplasts appeared completely free of wall remnants. After 30 min of regeneration, the plasmalemma of most protoplasts was covered by a thin net of apparently randomly arranged microfibrils. These microfibrils appeared tightly adpressed to the plasmalemma. Impressions of microfibrils in the plasmalemma were regularly observed both at the inner and outer fracture face of the plasmalemma. After 4 h of regeneration, the density of the microfibrillar net was much higher than after 30 min. The microfibrils often formed bundles even when the density of the microfibrillar net was still low. Besides the microfibrillar nets, the walls of regenerating protoplasts contained reticulate material that covered the microfibrils and often extended far into the medium. This material often formed thick irregular fibres. It is known that chitin and α-1, 3-glucan (alkali soluble S-glucan) are prominent in the walls at this stage of regeneration. The antibiotic polyoxin D prevents chitin synthesis but not α-1, 3-glucan synthesis (De Vries and Wessels, 1975). By using this drug, the appearance of wall components was studied. Regeneration in the presence of polyoxin D (100 μg/mL) resulted in the formation of walls which lacked the microfibrillar component but appeared to be entirely composed of reticulate material that must constitute the α-1, 3-glucan. Negative staining showed that this material actually consisted of thick irregular fibers (fibre diameter 20-30 nm). Similar fibers were observed in preparation of reprecipitated S-glucan from normal hyphal walls.

f. Examination of Shadowed Wall Preparations of Regenerating Protoplasts

Phase-contrast microscopy showed that walls isolated after 3 h of regeneration were very fragile and easily fell apart but walls prepared after 9 h of regeneration mostly retained the shape of the cell. Electron microscopy of shadowed preparations revealed that after 3 h of regeneration, most walls consisted of two components, i.e. a thin net of microfibrils and aggregates of amorphous material, which partly covered the microfibrils. Some wall fragments lacked the amorphous material. Treatment with alkali, which solubilizes α-1, 3-glucan, removed the amorphous material but did not change the appearance of the microfibrils. A minor fraction of wall fragments contained three components as shown by the fact that the inner and outer surfaces of the alkali resistant fragments were clearly different. This feature became prominent in wall fragments of primary cells isolated after 9 h of the regeneration. The outer surface of non-extracted wall fragments was rough and microfibrils were partly or completely covered by alkali soluble amorphous material. Alkali treatment completely uncovered the microfibrils at the outer surface of the wall of the primary cells. The inner surface of non-extracted walls was rather smooth revealing microfibrils which were embedded in an amorphous alkali-resistant matrix. Individual microfibrils often could be traced for considerable distance at the inner surface of the wall. Alkali treatment did not change the appearance of the inner wall surface.

The wall architecture of hyphae and hyphal initials of reverting protoplasts clearly differed from that of primary cells. The outer surface of these hyphal walls was smoother than that of the walls of primary cells and microfibrils. These walls also contained an outer cover of amorphous material, presumably α-1, 3-glucan. Removal of this cover with alkali revealed a very smooth surface without any visible microfibrils. There was sharp demarcation in wall architecture at the site of hyphal tube emergence.At this site, the microfibrillar net of the primary cell appeared to be ruptured and remnants of this net were sometimes observed on the surface of the hyphal tube initials. The inner surface of the hyphal walls was similar to that of the primary cells and buds. Its appearance did not change after alkali treatment: microfibrils were embedded in an alkali-resistant matrix. Treatment of alkali resistant wall fragments with hot dilute HCl or exo-β-1, 3-glucanase completely removed the matrix of both primary cells and hyphae. Together with its stainability with Thiery reagent, this suggests that the matrix is similar to the alkali-insoluble R-glucan (β-1, 3 and β-1, 6-glucan) of the normal wall and the microfibrils represent chitin, since chitin is resistant to these treatments.

De Vries and Wessels (1975) showed that alkali soluble S-glucan and chitin are the first wall components to be synthesized by regenerating protoplasts of S. *commune* whereas the synthesis of the alkali insoluble R-glucan lags behind (Fig. 14).

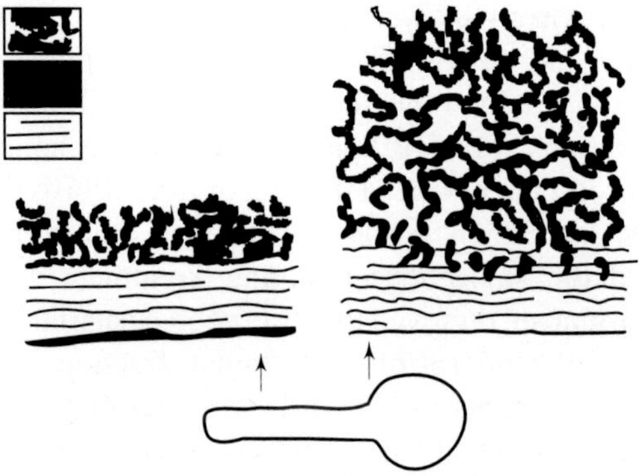

Fig. 14. Model of the ultrastructural location of wall polymers in primary cells and hyphae of reverted protoplasts of *Schizophyllum commune*.

Source: *Vander valk and Wessels (1976).*

g. X-Ray Diffraction of Isolated Wall Fractions of Regenerated Protoplasts (Wessels *et al.*, 1972)

Density tracings of X-ray powder of isolated wall fractions of regenerating protoplasts revealed the main reflections of S-glucan indicating the presence of microcrystalline α-1, 3-glucan in the native walls. The diffraction pattern of reprecipitated S-glucan prepared from these walls was identical to that of S-glucan isolated from normal hyphal walls. Treatment of the alkali-resistant wall fraction with exo-β-1, 3-glucanase did not change the diffraction pattern. The X-ray diffraction patterns of reprecipitated S-glucan (a) and 'native' S-glucan (b) walls from regenerated protoplasts cultured in the presence of polyoxin D (c) showed similarity between the patterns indicating that all the preparations contained microcrystalline S-glucan.

The distinct microfibrils of the wall represent chitin since, (1) they are resistant to alkali and dilute acid like authentic chitin, (2) X-ray diffraction spectra of preparations of these microfibrils showed typical reflections of chitin, and (3) such fibrils are absent from the walls when regeneration is carried out in the presence of polyoxin D, a specific chitin inhibitor.

h. Calcoflour Staining to Detect Regeneration

Calcoflour white has been used by biologists to localize cellulose, chitin and other glucans, because of its ability to form hydrogen bonds specifically withβ-1 andβ-1, 3 polysaccharides (Darkan, 1962; Dooijewaard-Kloorterziel, 1973; Haigler *et al.*, 1980; Kohno *et al.*, 1983; Sietsma and de Boer, 1973). Sietsma and De Boer (1973) and Kohno *et al.*, (1983) demonstrated with calcoflour white stain that freshly prepared protoplasts, respectively, *Pythium* sp. and *Alternaria kikuchiana* did not fluoresce at all, but after 1-2 h incubation in the regenerating medium most of them showed weak fluorescence that rapidly became brighter on prolonged incubation. Cell walls were regenerated from the original protoplasts prior to budding and also at the apex of control hyphae which developed later. Electron microscopy revealed that cell wall materials regenerated around the protoplasts, especially at 50 ppm, were loosely aggregated and thus dissimilar to those of intact hyphae, which usually consisted of compact inner and outer layers. Thus calcoflour white may be specific to loosely woven wall materials rather than to well-organized, intact cell walls. This speculation is supported by the fact that intact mycelium show weak fluorescence by the calcoflour white stain. Calcoflour staining is well suited to observe protoplast regeneration in biotrophs.

i. Autoradiography to Detect Regeneration

Autoradiography after a pulse of radioactive wall precursor exhibits the areas on the regeneration structures of wall synthesis. The protoplasts of

S. commune markedly differs in the pattern of incorporation of N-(acetyl 3)-glucosamine into the wall. The light microscope observations of autoradiography reveal the abnormally developing chain cell and also normal reversion. These differences in the pattern of incorporation of wall precursor such as the failure of some protoplasts to revert to normal growth has to do with the failure to organize local growth centre, where controlled extensions occur. In the absence of such centres, growth of the regenerating structure may only be possible by occasional, blow out, of the cell wall at a weak spot resulting in chain cells. This observation was confirmed by using chitin inhibitors (polyoxin D) with simultaneous inhibition of β-glucan. The regenerating walls containing mainly α-glucan develop internal pressure as a result of which, the regenerating structure simply increased in size without the formation of typical chain cells nor hyphal tubes (De Vries and Wessels, 1975).

2.3 BIOCHEMICAL ASPECTS OF WALL BIOGENESIS

a. Wall Biogenesis in Yeast Fungi

Chemical and physical analyses on the nets of *Saccharomyces* spp. have given conflicting results. The first analysis, made on *S. carlsbergensis*, revealed the presence of mannose, glucose and N-acetyl-glucosamine and traces of amino acids. The hexosamine content was high at 27%, which suggests a high chitin content. Kopecka *et al.,* (1974) reported that the fibrillar nets from *S. cerevisiae* were soluble in dilute alkali, indicating that they consisted of a glucan. Confirmation came from analyses on the hydrolysis products of purified nets, which revealed only glucose to be present. The latest observations continue to fuel the controversy. X-ray diffraction patterns on the regenerated wall, by using preparations free of contamination with wall remnants and bud scars, showed the presence of two microcrystalline components, an unbranched β-1, 3-glucan and chitin. The glucan formed the alkali soluble fraction, differing from the normal yeast wall glucan in that β-1, 3-linkages were absent. The chitin component was quite light, which accounted for 10% of the net material, but its presence could not be demonstrated in electron micrographs. An earlier report on the regenerated protoplast wall in *Candida utilis* also suggested the presence of a high chitin content.

The controversy over the presence of chitin resides in the possibility of contamination of the material with bud scar residues. What may be significant in this context is the uncoupling of nuclear division and septum formation in the early stages of protoplast reversion. It is possible that pulses of chitin synthesis could occur following nuclear division, as in budding cells, but in protoplasts the polymer ends up in the fabric of the wall due to a possible change in membrane organization.

The ability of protoplasts from fission yeasts and filamentous fungi to undergo reversion in liquid medium can be readily understood with respect to the ultrastructural differences that occur in comparison to the budding yeasts. Protoplasts of *Schizosaccharomyces pombe* produce a microfibrillar net that increased in density as regeneration proceeds, and later an amorphous component becomes associated with it. The size and density of the fibrils have been regarded as significant in the retention of other wall component, and in solid media the fibrillar and amorphous components are laid down simultaneously. Recent work has demonstrated that the fibrillar net is comprised of a highly crystalline β-1, 3-linked glucan, with the subsequent incorporation of an α-1, 3-glucan as regeneration proceeds.

b. Wall Biogenesis in Filamentous Fungi

Wall regeneration by protoplasts from mycelial fungi resembles the situation in yeast in several respects. The process commences with the deposition of a microfibrillar component forming a network that encloses the protoplast, and later an amorphous component is deposited within the network. In some instances, the network is produced very rapidly, in as little as 10 min.

The chemical composition of newly regenerated wall has been examined in several fungi. In *Rhizopus nigricans* the amorphous component could be solubilized with trypsin, leaving the microfibrillar material that was removed by chitinase. A chitinous microfibrillar network was also found in *Aspergillus nidulans,* together with a amorphous component that is yet to be identified.

In the early stages of wall regeneration in *Pythium* the protoplast forms a loose network of fibrils, which becomes dense and compact with time. This contrasts with *S. commune* where a layering of wall material was found. Initially, the protoplasts produce a loosely organized fibrillar net of chitin, which later becomes covered by a fluffy glucan layer. As wall regeneration continues, the third polymer, β-1, 3-glucan, was deposited as a matrix in the interstices of the chitin network. The outer fibrils of the network remain free and covered only by α-glucan forms a compact layer. Therefore these differences in architecture could be more important as determinants of hyphal morphogenesis of wall composition.

In only a few cases have the ultrastructural observations been correlated with chemical analyses on the newly formed wall material. The paucity of information is a reflection of several technical problems associated with dealing with relatively small amounts of material and with systems that lack synchrony in development. Where investigations have been made the latter does not seem to have been such a problem.

c. Synthesis of cell-wall polymers

The most comprehensive study involved *S. commune* with chemical analysis and isotope incorporation methods. A clear pattern of polymer deposition was found, that correlated with the ultrastructural studies. At the start of regeneration, chitin and α-glucan were produced followed by β-glucan after a few hours lag. Reversion, i.e. germ tube development, did not begin until polymer deposition had been taking place for several hours. With the inhibitors polyoxin D and cycloheximide, it proved possible to uncouple the biosynthesis of the three polymers. Both chitin and β-glucan synthesis were blocked in presence of polyoxin D, leaving α-glucan to be formed as thick loose fibers. With cycloheximide, chitin and α-glucan synthesis occurred as normal for the first 5 h and were inhibited. β-glucan synthesis did not occur. These experiments suggested that chitin synthase and α-glucan synthase were present in the protoplasts at release, but β-glucan synthase had to be formed *de novo*. β-glucan synthase may be a wall bound enzyme in the intact hyphae and would be lost at protoplast release.

The three main cell wall polysaccharides namely chitin, S-glucan and R-glucan were synthesised during wall formation. The amount of S-glucan and chitin apparently increased almost linearly from the start of regeneration. The accumulation of R-glucan showed a lag of about 3 h. After this period, R-glucan accumulation was roughly linear for about 8 h and occurred faster than S-glucan accumulation. Hyphal tubes started to emerge after the accumulation of R-glucan (De Vries and Wessels, 1975).

d. Protein and Nucleic Acid Syntheses

The total amount of protein and nucleic acids remained fairly constant during the first hours of regeneration after which there was a small but definite increase. On the other hand, the early labelling of pronase digest suggests that active protein synthesis occurred even during the early stages of regeneration. Indeed continuous labelling with ^{14}C-adenine and ^{14}C-leucine showed that active nucleic acid and protein syntheses did occur. The incorporation of adenine and leucine was roughly linear, starting 30 to 60 min after the onset of the regeneration, through at least 7 h. Pulse labelling revealed that at the onset of regeneration the incorporation rate of both adenine and leucine was very low. For both compounds, the rate of incorporation increased steeply during the first 90 min to reach and almost constant level. After 7 h of incubation, the incorporation for adenine increased again, whereas incorporation rate of leucine dropped after 5 h of incubation.

e. Effect of Cycloheximide on Protein Synthesis

The effectiveness of cycloheximide to inhibit protein synthesis in regenerating protoplasts of *S. commune* was investigated by measuring its effect at various concentrations on the incorporation of ^{14}C-leucine into TCA precipitable material. The concentration of cycloheximide at 0.5 μg/mL caused complete inhibition of leucine incorporation. However, the percentage of regeneration, determined after 3 h incubation period, was not affected at any concentration of the drug.

In a time course analysis of the formation of the three main cell wall components by protoplast regeneration in the absence and presence of 0.5 μg/mL cycloheximide, S-glucan and chitin accumulated apparently directly from the start and a delay of about 3 h was found for R-glucan accumulation. Emergence of hyphal tubes began after about 8 h. In the presence of cycloheximide, S-glucan nor chitin accumulation appeared to be affected initially, but after 5 h the synthesis of these compounds decreased. Accumulation of R-glucan was strongly inhibited and virtually no reversion to hyphal growth occurred. These results indicate that the cell's machinery for synthesizing S-glucan and chitin was still present in the isolated protoplasts, whereas the deposition of R-glucan required *de novo* protein synthesis.

f. Effect of Polyoxin D on Chitin Synthesis

Polyoxin D strongly inhibited the formation of chitin and at concentrations higher than 25 μg/mL chitin was almost absent. Surprisingly, the deposition of alkali-insoluble glucan was even more affected by the drug and virtually no R-glucan was found even at the lowest level of polyoxin D tested. The synthesis of alkali-soluble/acid precipitable glucan was not inhibited nor stimulated.

In polyoxin D treated cultures no significant differences were found between polysaccharides obtained by fractionation of total TCA precipitates or of sediments obtained after osmotic shock. Clearly all polysaccharides belonged to the cells and did not include extracellular materials. Treatment of the S-glucan fractions with crude 'R-glucanase' for 20 h released less than 10% of the anthrone positive material which indicates that the major part of S-glucan was α-1, 3-glucan. In addition, the supernatant of S-glucan precipitate derived from cells after osmotic shock contained a very small amount of anthrone positive material. Walls regenerated in presence of polyoxin D are practically devoid of β-glucan and consisted mainly of α-1, 3-glucan.

In the presence of polyoxin D, the reversion to hyphal growth was completely inhibited, even at the lowest concentration of polyoxin D used. Clumping of cells was observed, but the clumps were less compact and more easily dispersed. In contrast to normal regeneration, the cells and particularly their vacuoles, increased in size despite the presence of a wall. These walls were not mechanically weak since the cells were osmotically stable. Occasionally wall structures with a more or less tubular morphology were seen, but their diameter exceeded that of normal hyphae. Particularly after prolonged incubation spherical, optically empty structures were observed. These structures were no longer detected after osmotic shock and therefore they probably represented isolated vacuoles remaining from disintegrated cells.

Polyoxin D did not prevent the occurrence of nuclear division since in most of the cells, numerous nuclei could be readily observed. Measurements of oxygen uptake showed that polyoxin D caused little effect on the respiratory activity of protoplasts.

2.4 PROTOPLAST REGENERATION AND REVERSION IN SOME FILAMENTOUS FUNGI

Regeneration on of protoplasts differ in different types of fungi (Fig. 15).

a. *Neurospora crassa* (Emerson & Emerson, 1958)

Emerson and Emerson (1958) described reversion of *Neurospora* sp. protoplasts upon transfer to media lacking the lytic enzymes (Snail lytic enzyme) used for the digestion of cell wall. Complete reversion to typical mycelial growth required from a few hours to several days and the type of mycelium formed seemed to depend upon the osmotic strength of the medium. Similar observations were made by Bachmann and Bonner (1959) where protoplasts of *Neurospora crassa* obtained with the snail lytic enzyme system, when transferred to a suitable liquid or solid nutrient medium regenerated to give normal mycelial growth. The course of regeneration varied greatly.

b. *Bipolaris oryzae* (Annamalai, 1991)

Protoplasts had the potential to regenerate on media amended with osmotic stabilizers (CZYAM and PDA amended with 0.6 M sucrose-mannitol) had higher per cent of regenerated protoplasts than the other media amended with osmotic stabilizers. After 24 h, 25% of the protoplast regenerated on CZYAM media containing 0.6 M sucrose-mannitol. At 48 h, the per cent of regenerated colonies increased to 48%, but further incubation did not result in the increase of regenerated colonies. The protoplasts incubated on water agar with 0.6 M mannitol, showed only 8% of regeneration and produced viable colonies after 48 and 72 h. Protoplast did not regenerate on media without osmotic stabilizers. Regenerating protoplasts of *B. oryzae* directly produced germutbes in all media containing osmotic stabilizers such as mannitol and sucrose (Plate 10c).

c. *Venturia inaequalis* (Ambri strain, Vijayapalani, 1995)

The various stages of regeneration process of *V. inaequalis* protoplasts (Plate 11) show no change in appearance. After 8 h the majority of the protoplasts developed a protrusion which increased in volume. Some protoplasts formed

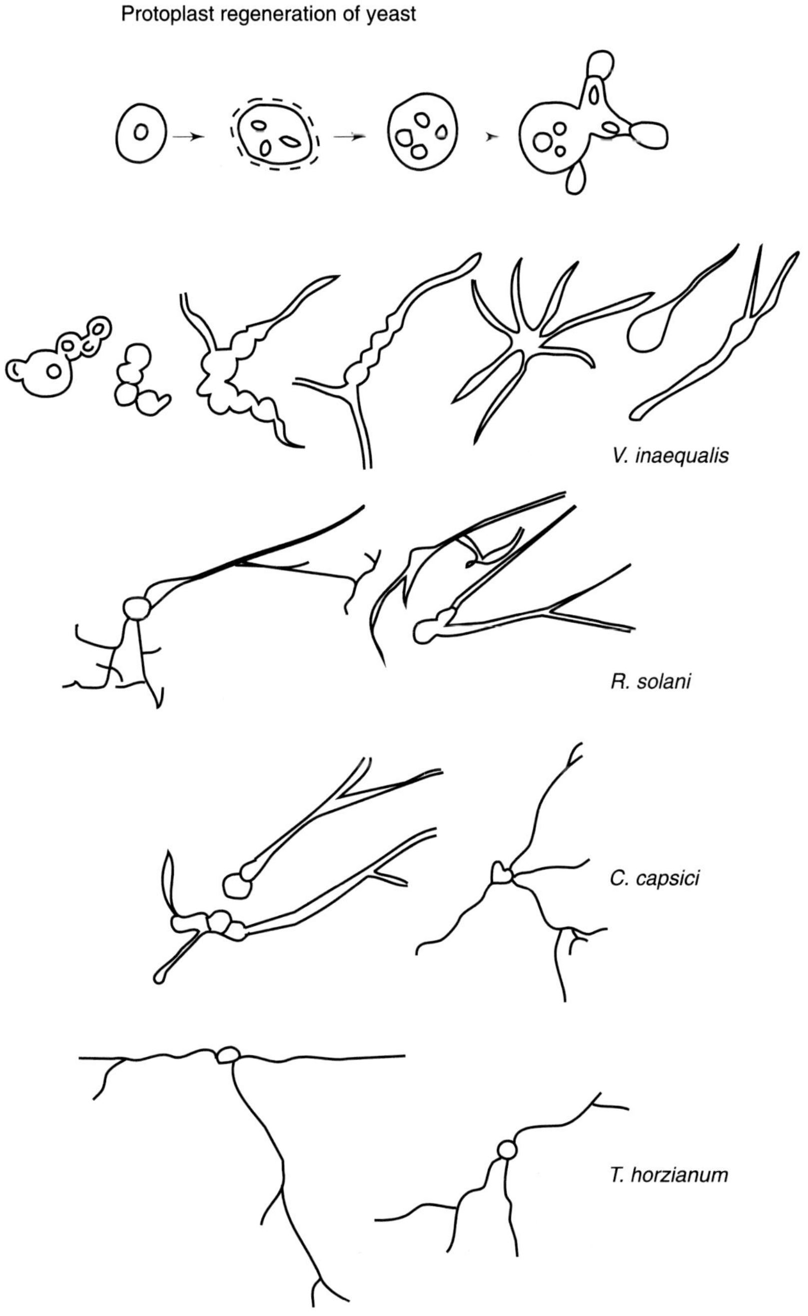

Fig. 15 Various types of regeneration of pp. in filamentous fungi

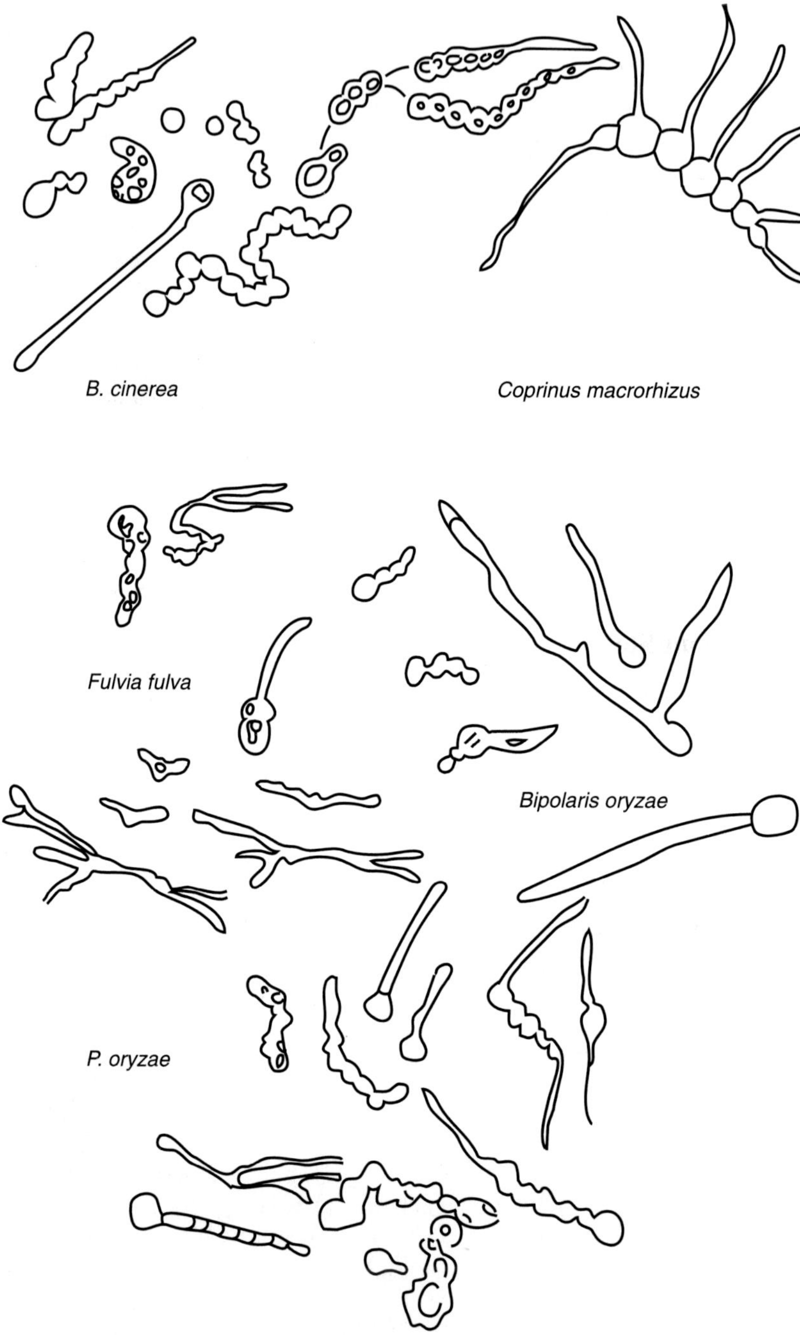

Fig. 15 Various types of regeneration of pp. in filamentous fungi

large vacuoles. Some protoplasts grew during the first 24 h, giving more or less pseudomycelial figures with a wide range of shapes. Other protoplasts developed a kind of spherical structure, not sensitive to osmotic shock, which was able to form hypha. *V. inaequalis* showed three different regeneration patterns of protoplasts. In the first type, a bud was formed from the regenerated mother cell from which a series of buds were formed. This type later did not revert to mycelium. In the second type, protoplast produced a bud from which a primary hyphae developed. In the third type, primary hyphae developed directly from the regenerated mother cell. In liquid medium only the first type of reversion was observed. The number of primary hypha varied from 1 to 10. The primary hypha lengthened, branched and ultimately reverted to mycelium. On continued incubation, these colonies showed the typical characteristics of the parent colony and ultimately produced conidia. (a) The protoplasts gave rise to a series of buds, simulated yeast-like growth and stopped there off. They did not form colonies. Some of these forms differentiated, more frequently than their neighbours and finally produced a germtube capable of more or less normal hypha-like growth. (b) In the second type of regeneration, the protoplast itself germinated to produce a hyphal tube. Individual protoplasts produced one or more hyphae as direct outgrowths of the mother cell. (c) In the third type the original protoplasts, keeping their spherical shape, produced a protrusion which enlarged, giving another spherical body from which normal hypha developed and formed microcolonies. Some protoplasts developed another protrusion which enlarged into another spherical cell (secondary). The content of the primary cell emptied into the secondary cell and the shells of the primary protoplasts is called Ghost cells.

The remains of the shells (cell membrane or cell walls) can be seen in the preparation when the protoplasts grew to a mass of more or less bud-like and spherical forms, some with large refringent vacuoles and some forming germtubes to give normal mycelia. It was also observed that the empty shells were sometimes filled again, so that the cytoplasmic material moved from one side to another leaving empty spaces during this transfer. The meaning of this phenomenon is unknown.

Prolonged incubation of the first generation of globular forms from *V. inaequalis* protoplasts invariably resulted in mass reversion to the mycelial state. This occurred through an initial increase in the number of large bodies following their separation and in the formation of enlarged filamentous forms. It should be emphasized that all the reversion studies were made of semi-solid medium, which is a more effective means of achieving reversion than on solid or liquid media.

d. *Trichoderma spp* (Mrinalini, 1997)

A known aliquot (0.1 ml) of the isolated protoplasts of *T.harzianum* was aseptically spread on various regeneration media namely water agar, potato dextrose agar, potato dextrose agar (PDA) amended with osmotic stabilizer (0.6 M KCl), PDA amended with bavistin (0.1 μM) and osmotic stabilizer (0.6 M KCl). Considerable variation in regeneration was observed between different regeneration media. Regeneration frequency reached 56% when PDA with 0.6 M KCl was used as the regeneration medium. There was no regeneration in PDA without osmotic stabilizer. There was no regeneration at all in water agar and PDA with bavistin amended medium. It was observed that the regenerating protoplast initially produced a bulbous projection which later on resolved into hyphal form of growth. This typical observation was made on both PDA and PDA amended with osmotic stabilizer (0.6 M KCl). *T. longibrachiatum* also showed 48% regeneration frequency after 48 h of incubation. In both *T. harzianum* and *T. longibrachiatum* only one type of regeneration was observed (Plate 12).

e. *Rhizoctonia solani* (Elvarasan, 1996)

In the case of *R. solani* two different types of regeneration were observed (Plate 13).

Protoplasts directly gave rise to thin hyphae with rapid branching to form Plate-13 colony. The second type of regeneration was initiated by bud-like formation subsequently producing thin mycelia. Budding form of regeneration was not observed in *R. solani.* More than one germ tube was produced by a single protoplast indicating the multinucleate condition of the protoplast. In both the types of regeneration microcolonies were produced.

f. *Fusarium culmorum* (Garcia *et al.*, 1966)

Majority of the protoplasts began to develop a protrusion which increased in volume after 8 h of incubation. The earliest visible alteration in the morphology of the protoplasts was the formation of cellular aggregates. Some protoplasts divided and grew during the first 24 h, giving more or less pseudomycelial figures with a wide range of shapes. Other protoplasts developed a kind of spherical structure, not sensitive to osmotic shock, which formed hyphae. Three different regeneration patterns of protoplasts were observed.

a. The protoplast gives rise to a series of yeast-like forms grouped in a chain. Some of these forms differentiated, becoming more refringent than their neighbours and finally producing a germtube capable of more or less normal hyphae-like growth.
b. The protoplast itself germinated to give a hyphal tube, but this was very uncommon. Individual protoplasts sometimes produced one or more hyphae as direct outgrowths of the mother cell.

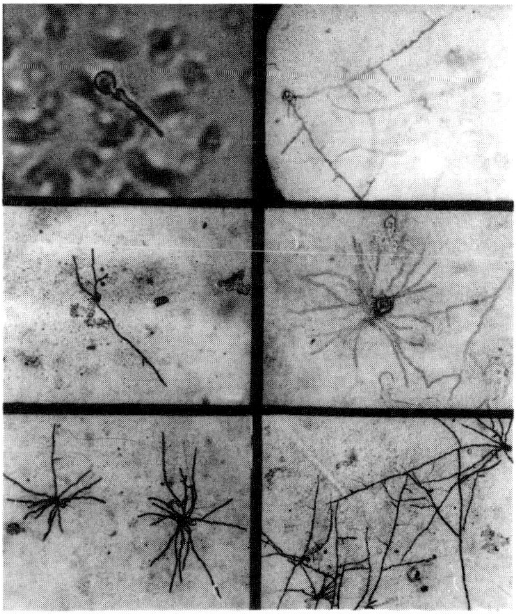

Plate 11. Emergence of a primary hypha from a (uninucleate) bud (960×).
Direct emergence of primary hyphae (binucleate).
Emergence of three primary hyphae (trinucleate) (350×).
Emergence of more than three primary (multinucleate) hyphae (400×).
Microcolonies after 36 h (160×).
Microcolonies after 72 h (160×).

Source: *Vijayapalani, 1995.* Ph.D. Thesis, University of Madras, India, 97-225.

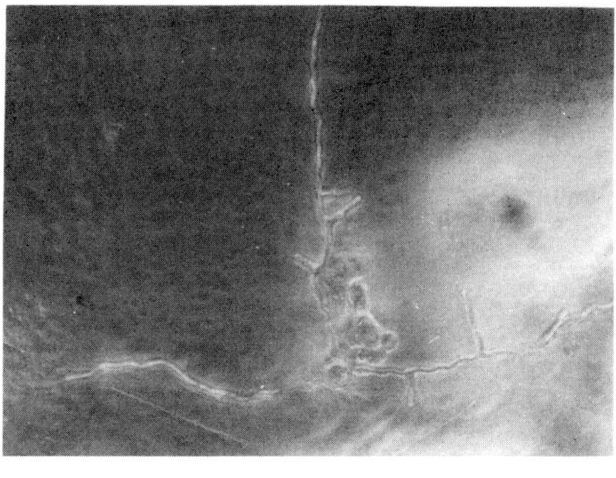

Plate 12. Regeneration of protoplasts of *Trichoderma* spp.

Source: *Mrinalini, 1997.* Ph.D. Thesis, University of Madras, India, 66-102

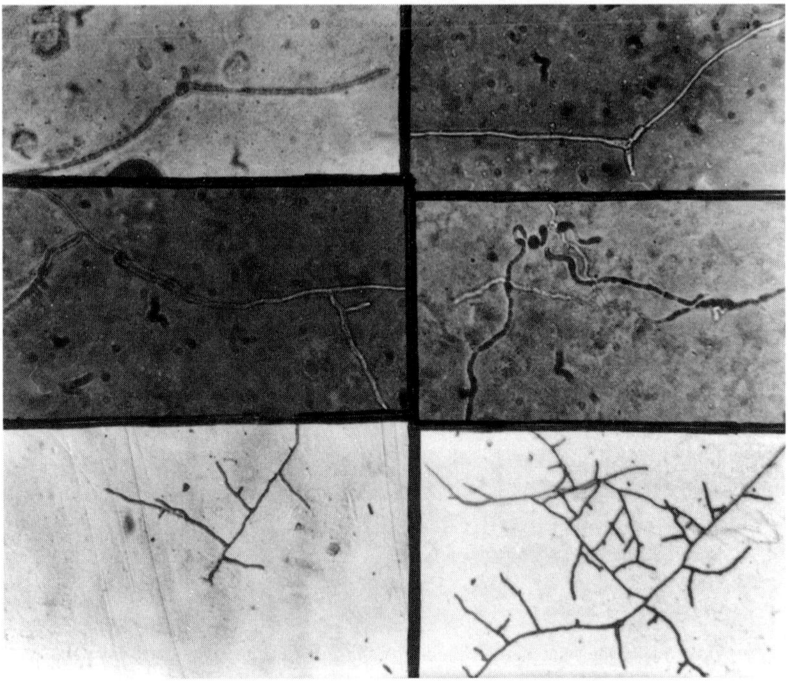

Plate 13. Regeneration of *R. solani* protoplasts.
Source: Elavarasan, 1996. Ph.D. Thesis, University of Madras, India, 96-97.

c. The original protoplast, keeping its spherical shape, produced a protrusion which enlarged, giving another spherical body which remained beside the first. Sometimes one of these bodies gave another amoeboid form and the contents of the previous one were transferred to the neighbouring cell.

The remains of these shells (cell wall) can be seen in the preparation when the protoplasts grow to a mass or more or less bud like and spherical forms, some with large refringent vacuoles and some forming germ tubes to give normal mycelium. Mycelium originated from an isolated protoplast, after transfer to fresh nutrient medium was able to sporulate normally. These reversion cultures were indistinguishable from the original culture in their growth habit, morphology, chemical composition, sensitivity to strepzyme RA and productivity of protoplasts. This means that a normal mycelium can originate from a protoplasts absolutely devoid of cell wall. There is no doubt that the mycelial protoplast from *F. culmorum* is a true naked protoplast since the cell wall remains in the preparation, empty but keeping its shape, once the protoplast has been released. Electron microscopic studies of ultra thin sections of naked *Fusarium* protoplasts showed no remnants of cell wall on the surface of the cytoplasmic membrane. As a consequence it can be concluded that on regeneration these protoplasts develop thick cell walls suggesting *de novo* synthesis of rigid envelope. The presence of some vesicles observed in the inner side of the plasma membrane of ultra thin protoplast sections suggested that those structures may play an important role as carrier of enzymes or precursors of the wall by a process equivalent to reverse pinocytosis (Villanueva, 1966).

g. *Ophiostoma ulmi* (Muthukumar *et al.*, 1985)

Spores and mycelial protoplasts regenerate into actively growing fungal colonies when incubated on agar medium. Purified spore protoplasts were suspended in 2.4% PD broth prepared in MCE medium and diluted samples were overlayered onto the surface of 1% Bacto-agar. Incubation was at 25°C and growth of regenerating protoplasts was observed during a 48 h period by phase contrast microscopy.

After 2 h of incubation, single bud-like structures, appear on the surface of the protoplast. After 3 h of incubation, these bud-like structures multiplied as short chains and later developed into a germ tube. After 20 h of incubation, some protoplasts developed a single germtube which developed into a primary hypha.

Some protoplasts developed two germ tubes which developed into two hyphal tubes. Similar development was also observed with regenerating spore protoplasts. The similarity in growth and development that was observed for regenerating spore and mycelial protoplasts may be the result of the morphological commitment phenomenon for *O. ulmi*, previously reported by Muthukumar *et al.* (1985). Protoplasts liberated from *Rhizoctonia solani* (Hashiba and Yamada, 1982) had also been reported to regenerate in a similar fashion to that reported here for *O. ulmi*. After 48 h of incubation, approximately 50% of the spore and mycelial protoplasts regenerated into actively growing fungal colonies.

h. *Geotrichum candidum* (Fukui *et al.*, 1969)

Geotrichum candidum protoplasts incubated in Basal Growth Medium (BGM) at 30°C germinated after 6 h and divided into an original and first generation by a septum after 7.5 h. After 8 h the second germtube usually appeared at right angles to the first, and after 17 to 20 h, the single protoplast developed into a colony. The process of regeneration of protoplasts of *G. candidum* has been divided into three phases viz. lag, logarithmic and germinating phases. When protoplasts were placed in stabilizing BGM, they became slightly swollen during the 2 h lag period. Germinating cells first became visible after 2 to 3 h of incubation, but they were poorly resistant to distilled water. The maximum number of osmotically stable cells was seen after 5 h of incubation, whereas osmotically stable germinating cells represented about 18% of the total protoplasts at that time. The number of osmotically stable, germinating cells rapidly increased in the next hour and represented approximately 100% of the total protoplasts after 7 h.

After 24 h a few protoplasts showed irregular growth forms similar to those reported by Gascon *et al.* (1965) for various yeasts and mold protoplasts. However, the most abnormal cells lyse within an incubation period of 24 h.

i. *Pythium* PRL2142

Pythium protoplasts regenerate into new mycelium when suspended in a medium with suitable osmotic protection and adequate nutrients. After 3 h of incubation some protoplasts started to form one or more buds, which grew out as normal hyphae. Some protoplasts extended to a structure with several swellings before a normal hypha arose. However, a great number of protoplasts were still unchanged after 18 h of incubation. The regeneration process was followed under fluorescence microscope after incubating samples with calcofluor (an optical brightner). This substance adheres specifically to glucans and make them fluoresce strongly when illuminated with near UV light (Preece, 1971). Freshly prepared protoplasts did not fluoresce at all, but after 1 h of incubation in the regeneration medium most of them showed a weak fluorescence, that rapidly became brighter on longer incubation. Newly formed buds and hyphae fluoresced very strongly. The tip of the hyphae generally fluoresced more strongly than the rest of the mycelium.

j. *Pyricularia oryzae* (Asai *et al.*, 1986 and Lalithakumari and Saradha Kumari)

Protoplasts from spores and hyphae of *P. oryzae* had the potential to regenerate and reverse. About 50% of spore protoplasts and about 90% hyphal protoplasts germinated when incubated in Czapek yeast liquid medium (CZYLM). The germination was hyphal and budding type. In hyphal type, normal shaped hyphae were reversed directly from the regenerated protoplasts and in budding type budded cells were formed. There is a third type where normal hyphae sometimes arise from apical budded cells (Fig. 13). Budding type some time stop growth and disintegrate.

k. *Saccharomyces cerevisiae* (Anne, 1993)

Protoplasts of *Saccharomyces cerevisiae* and a few other yeasts stand out in comparison with all other fungal protoplasts in that reversion to normal yeast cells occurs only in solid media (Necas, 1971; Svoboda, 1966 and Svoboda and Necas, 1966). In liquid medium, the protoplasts form an incomplete wall that prevents complete reversion. Reversion frequencies for yeasts cultured in gelatin or agar-solidified media are 50-70% for *S. cerevisiae* (Svoboda,

1966) and 90% for *S. pombe* (Havelkova, 1969). Protoplasts of *S. cerevisiae* in solid medium maintain their spherical form during the early period of new wall formation (Necas, 1971). Biosynthesis of wall materials is accompanied by a size increase and nuclear division. Removal of the cell wall during protoplast formation apparently blocks the synchrony between nuclear divi sion and septum development. The first generation of buds to arise from this multinucleate cell are also atypical in shape and may also contain several nuclei. The normal elliptical form of the *S. cerevisiae* cell is found at the second generation bud. In contrast, the septation of nuclear division and cytokinesis described for *S. cerevisiae* is not found in *S. pombe.* The wall forming protoplasts retain spherical form and at the first nuclear division two hemispherical cells are formed following septation which later develops into typical *S. pombe* elongated cells.

l. *Hebeloma cylindrosporum* (mycorrhizal fungi) (Hebraud and Fevre, 1987)

Mannitol and sorbitol (0.7 M) allowed regeneration of *H. cylindrosporum.* Two classical regeneration of reversion were observed. Some protoplasts produced yeast-like buds that developed into a chain of irregular cells and later into hyphae which resembled germtubes. Enlarged protoplasts developed directly into germtubes in one or more directions. The reversion to colonies was very slow, and were distinguishable only after one week incubation. The regeneration rate of protoplasts was similar to other Basidiomycetes but higher to that of another ectomycorrhizal fungus (Kropp and Fortin, 1986).

m. *Penicillium chrysogenum* (Anne, 1977)

In a suspension of purified protoplasts, treated with 0.1% Tinopal BOP for 5 min and viewed under a fluorescence microscope, between 85 and 95% of the total number of protoplasts were completely free of fluorescence. In contrast, mycelial walls and septa stained very intensively and protoplasts in hypertonic production medium (PM) showed weak fluorescence after 1 to 2 h of incubation, indicating that they had rebuilt some cell wall material. On further incubation (2-4 h), protoplasts looked brighter and 3 types of reversions were observed in *P. chrysogenum.* A small number of protoplasts formed a bud at one side. This protrusion grew as large as the protoplast itself and formed another bud giving rise to a short chain of yeast-like cells. This chain always remained very short and was composed of 4 cells at the most, contrary to the reversion pattern of other species including *C. acremonium, F.culmorum* (Garcia *et al.,* 1966) and *A. nidulans* (Peberdy and Gibson, 1971), which formed chains of up to 20 cells. One or more of these cells formed a germtube capable of hypha-like growth. In other cases, a bud on the protoplast deve- loped into a convoluted hyphal thread on incubation. Most frequently, enlarged protoplasts directly developed germtubes in one or more directions from the protoplasts. These hyphae branched and formed a microcolony after 8-12 h of incubation. The choic of the stabilizer and liquid or solid medium influenced to some extent the proportion of reversion patterns.

The number of nuclei in the protoplast could influence the reversion pattern. Furthermore, types of protoplast reversion were morphologically different from the development of germtubes in spores.

n. *Colletotrichum capsici* (Kalpana, 1995)

Regeneration and reversion of isolated protoplasts using PD agar and water agar amended with different osmotic stabilizers, i.e. NaCl (0.6 M) and KCl (0.6 M) and incubated for 24 h, revealed that potato dextrose agar with 0.6 M NaCl was the best medium for regeneration and reversion of protoplasts of *C. capsici*.

The regeneration of protoplasts was carefully followed and photographed using phase-contrast microscopes at different developmental stages. Each protoplast produced more number of germ tubes. The regeneration tube was as normal as that of the conidial germination (Plate 14). Each protoplast colony was isolated and maintained as individual culture. Protoplasts produced thin germtube on incubation without budding. Two types of regeneration were observed in *C. capsici* protoplasts. Budding observed in few cases was followed by normal mycelium. In the second type, mycelial form developed directly from the protoplasts.

a. Variations in the growth kinetics of regenerated protoplast isolates

a. Trichoderma harzianum (Karpagam, 1994)

Among the 14 protoplast isolates, variation in growth rate in liquid medium (PD broth) was observed. The isolates P1 to P7 and P9 showed a rapid growth within 24 h of incubation over control (wild or source or parent strain). The remaining isolates showed a slow rate of growth for the first 24 h. The growth pattern was changed in completion of 48 h of incubation, except P9, the other isolates showed rapid growth. To add further, at 96 h of incubation, all the protoplast isolates showed more growth than the parent strain. Among the 14 isolates tested P10 was observed to be more virulent in its growth rate reaching a maximum of 1,530 mg dry wt in 96 h than all the other isolates.

b. Colletotrichum capsici (Kalpana, 1995)

The growth rate of *Colletotrichum capsici* in terms of dry weight estimation reached the maximum growth on the 3 d of incubation. The parent strain reached the maximum growth on the 5 d of incubation and thereafter showed a

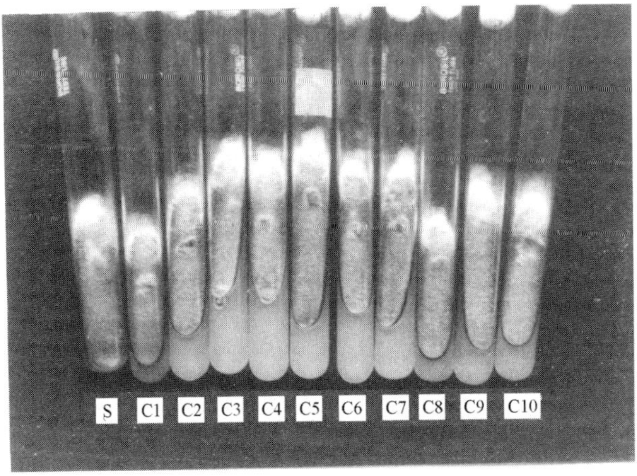

Regenerated protoplast cultures

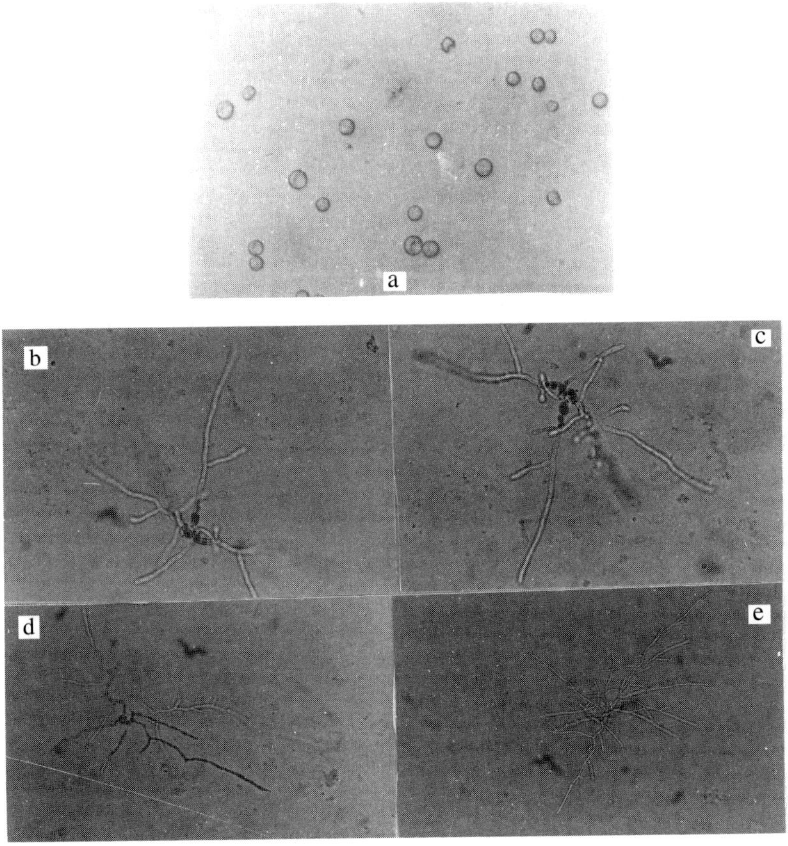

Plate 14. *C. capsici*. a: Isolated protoplasts, b, c, d and e; Regeneration of protoplasts.
Source: *Kalpana, 1995*. M. Phil, Thesis, University of Madras, India, 26-34.

steady decline. The growth rate in general in protoplast cultures C3, C6, C7 and C9 were slower and less than the other protoplast cultures. Among the 10 protoplast cultures C4 and C10 showed a rapid rate of growth and the increase in growth was marked from the first day of incubation.

b. Variation in morphology and pigmentation of regenerated protoplast cultures

The morphology of *C. capsici* regenerated protoplast culture grown on PDA plates show variations (Colour Plate 2a).

The protoplast culture C2 was light brown with reduced conidiation. The protoplast culture C3 showed typical concentric zonation with light and dark regions, though such zonations were seen in C4, C6, C7, C8 and C10, they were not as marked as they were in C3. Sectoring was observed in C4 and C8 like that of the source culture.

c. Variation of Growth on Chitin Amended Medium

The growth rate of the regenerated protoplast isolates of *T. harzianum* on chitin amended medium incubated for 120 h, (Colour Plate 2b) was slow in all the isolates except for the P10 over control for 24 h incubation but on 96 and 120 h incubation, there is no marked variation in the growth rate between the protoplast isolates and source culture. Sporulation was more in many protoplast isolates namely P2 and P4 to P14 of *T. harzianum* though maximum sporulation was observed with P10 isolate while a sweeping reduction in sporulation was observed in the isolate P3. Another interesting observation is the loss of green pigmentation. Change in pigmentation was observed in all the isolates grown on chitin amended medium.

d. Variation of Growth on Cellulose Amended Medium

The growth rate of regenerated protoplasts of *T. harzianum* on cellulose amended medium (Table 18), showed slow growth on 24 h of incubation.

Here again the P10 isolate showed a rapid growth when compared to other isolates. On further incubation, the protoplast isolates and source strain reached the maximum growth at 96 h. The spore production was more in many protoplast isolates namely P9, P11, P12, P13 and was very less in one isolate P6. The growth rate of some of the protoplast isolates were more in cellulose amended medium than chitin amended medium.

e. Variation of growth on pectin amended medium

On pectin amended medium (Table 19) again the regenerated protoplasts (P7 and P8) of *T. harzianum* showed a delayed growth.

Table 18 Growth kinetics of regenerated protoplast isolates on cellulose amended medium

Trichoderma harzianum	Radial growth in diameter (mm) Incubation time (h)				Spore concentration $\times 10^7$/ml
	24	48	72	96	
Source strain	16	34	69	90	72
P1	10	30	70	90	71
P2	12	31	71	90	70
P3	13	32	69	90	71
P4	12	34	71	90	72
P5	13	39	72	90	76
P6	12	30	67	90	39
P7	14	29	66	90	57
P8	15	29	67	90	78
P9	15	33	68	90	78
P10	23	64	90	90	73
P11	15	23	81	90	78
P12	18	24	68	90	87
P13	16	31	69	90	83
P14	15	32	71	90	72

Source: Karpagam, 1994. M. Phil. Thesis, Univ. of Madras, 30-60

Table 19 Growth kinetics of regenerated protoplast isolates on pectin amended medium

Trichoderma harzianum	Radial growth in diameter (mm) Incubation time (h)				Spore concentration $\times 10^7$ / ml
	24	48	72	96	
Source strain	15	50	80	90	73
P1	10	53	81	90	71
P2	18	63	90	90	71
P3	10	45	68	90	72
P4	9	44	68	90	73
P5	12	54	81	90	77
P6	9	41	65	90	40
P7	-	35	61	90	58
P8	-	35	60	90	60
P9	10	48	65	90	77
P10	21	70	90	90	73
P11	13	52	80	90	78
P12	-	38	63	90	86
P13	12	52	80	90	85
P14	9	41	66	90	74

Source: Karpagam, 1994. M. Phil. Thesis. Univ. of Madras–India 30-60.

The P10 isolate showed a rapid growth at 72 h. But all the isolates inclusive of source isolate showed maximum growth only at 96 h. Maximum sporulation was observed in protoplast isolate P11, P12, P13 and reduction in spore production was observed in the protoplast isolates P6 to P8. There was no other morphological changes in the protoplast regenerated isolates.

f. Total Protein Content of the Regenerated Protoplast Isolates

The total protein content of the protoplast isolates of *T. harzianum* and the source strain (Table 20) did not show distinct variation, though slight increase in protein content was observed in P9, P5, P2, P3, P10 and P8.

Following this the isolated proteins were subjected to electrophoresis using SDS polyacrylamide gel. In general, the electrophoretic pattern of protein did not show much variation though intensity of the bands were altered in some isolates.

Table 20 Total protein content of the regenerated protoplast isolates

Trichoderma harzianum	*Total protein (mg/g dry wt)
Source strain	1.1
P1	1.1
P2	1.3
P3	1.3
P4	1.2
P5	1.4
P6	1.1
P7	1.1
P8	1.3
P9	1.4
P10	1.5
P11	1.1
P12	1.1
P13	1.1
P14	1.0

* Mean of 4 replications
Incubation time - 72 h
Medium-PDA

g. Total DNA and DNA base composition of the regenerated protoplast cultures

Unlike protein, a distinct increase in the total DNA content was observed in P9, P10 and P12 followed by P11, P8, P6, P4, P14, P13. The GC% was calculated using UV absorbance technique (Table 21).

Table 21 Total DNA and DNA base composition of the regenerated protoplast isolates of *Trichoderma* (S)

Trichoderma harzianum	Total DNA (μg/g dry wt)	GC%
Source strain	1.8	52.6
P1	1.8	51.7
P2	1.6	52.6
P3	1.7	54.0
P4	2.1	53.2
P5	2.1	54.1
P6	2.2	53.4
P7	2.4	52.6
P8	2.3	53.1
P9	2.5	54.0
P10	2.5	54.4
P11	2.4	54.2
P12	2.5	54.0
P13	2.1	54.1
P14	2.9	52.6

Source: Karpagam, 1994. M. Phil Thesis, Univ. of Madras, India 32-59

h. Variation in Sensitivity of Regenerated Protoplast Cultures to Test Fungicides

The regenerated isolates of *T. harzianum* (Karpagam, 1994) when screened on different concentrations of iprodione showed variation in sensitivity (Table 22) at different concentrations. Among the protoplast regenerated isolates tested, P10 showed high tolerance to Iprodione. More than 40% of growth was observed even at 11 mM concentration of iprodione. On the other hand, the isolate P4 was observed to be highly sensitive. In general, all protoplast isolates except P4 were observed to be tolerant to iprodione (9 mM). The colour plates (C.P. 3) give a clear picture of the morphological variation, pigmentation, sectoring and growth pattern of the 14 protoplast isolates on Iprodione amended medium. Besides the inhibition variation, the morphological variation was also observed in protoplast isolates when grown on bavistin amended medium. In general the ring formation is common in *Trichoderma* spp. owing to day and night effect. But, the protoplasts regenerated isolates showed distinct additional ring formation. The 14 protoplast isolates checked for bavist sensitivity, P6 and P5 followed by P13 showed more growth than the source strain at 1.0 μM concentrate. At 1.5 μM concentration, P10 and P11 isolates showed good growth while all the other isolates were inhibited.

Iprodione at 10 mM concentration inhibited growth rate of protoplast regenerated isolates of *Colletrichum capsici*. The natural variation observed between regenerated protoplast culture could be exploited for natural selection of potential traits for crossing.

Table 22 Variation in sensitivity to Bavistin and Iprodione by regenerated protoplast cultures

Trichoderma harzianum	Sensitivity to fungicides	
	Bavistin (μM)	Iprodione (μ M)
C	18	25
P1	40(+122.2)	38(+52)
P2	21(+16.6)	41(+64)
P3	22(+22.2)	41(+64)
P4	23(+27.7)	43(+72)
P5	14(+22.2)	31(+24)
P6	40(+122.2)	34(+36)
P7	33(+83)	40(+60)
P8	21(+16.6)	39(+56)
P9	18(0)	20(-20)
P10	40(+122.2)	72(+188)
P11	19(+5.6)	42(+68)
P12	35(+94.4)	42(+68)
P13	27(+50)	42(+68)
P14	20(+11.1)	43(+)

Values in parenthesis shows % increase (+) or decrease (–) over source strain.
Mean of 4 replications
Incubation time : 72 hours
Medium: PDA

i. Variation in Biocontrol Efficiency of Regenerated Protoplast Isolates

The antagonistic potential of various regenerated protoplast cultures of *T. harzianum* and source strain were checked using dual culture technique of Huang and Hoes (1976). The pathogens used were *Rhizoctonia solani* causing sheath blight of rice and *Fusarium oxysporum* f. sp *lycopersici* causing wilt of tomato. The antagonistic potential of regenerated protoplast isolates of *T. harzianum* was more when compared to the source strain at the end of 72 h incubation. Though the initial growth was less than the source strain at the end of 72 h, they showed very high antagonistic potential.

j. Pathogenicity and virulence variations in regenerated protoplast cultures of *C. capsici*

Variations in virulence had been reported among protoplast isolates of *P. oryzae* by Hans *et al.* (1988) and in *Sclerotinia sclerotiorum* by Boland and

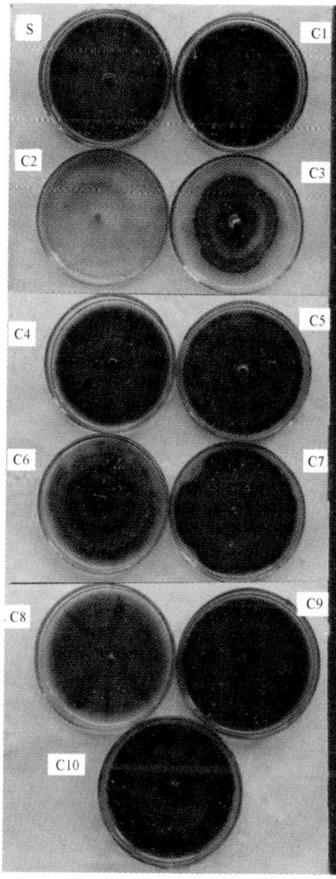

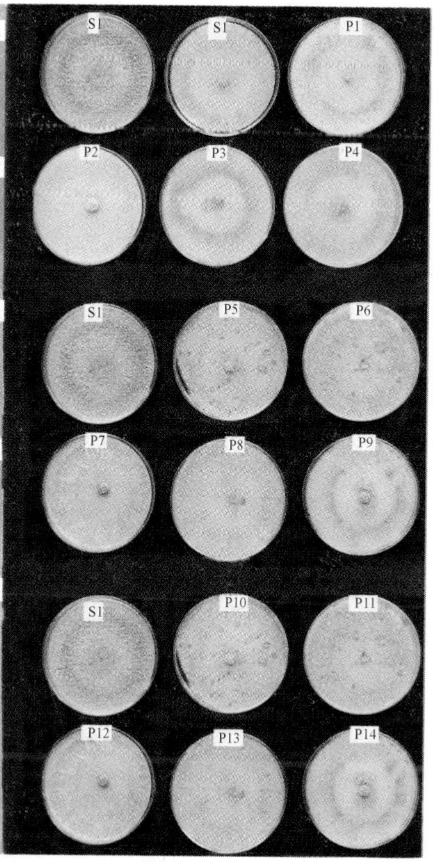

(a) (b)

Colour Plate 2a Morphology and pigmentation of regenerated protoplast cultures of C1-C10 of *C. capsici*
S-parent culture
Source: Kalpana, 1995.

Colour Plate 2b Growth kinetics of regenerated protoplast isolates of *Trichoderma* (P1-P14) on chitin amended medium.
S-parent culture on PDA medium
S1-Parent culture on chitin medium
Source: *Karpagam, 1994*. M. Phil Disserlation University of Madras, India, 30-57.

Smith (1991). Pathogenicity of the regenerated protoplast cultures of *C. capsici* was tested and compared with the source strain. The intensity was measured in terms of lesion diameter on the inoculated chilly fruits (Table 23).

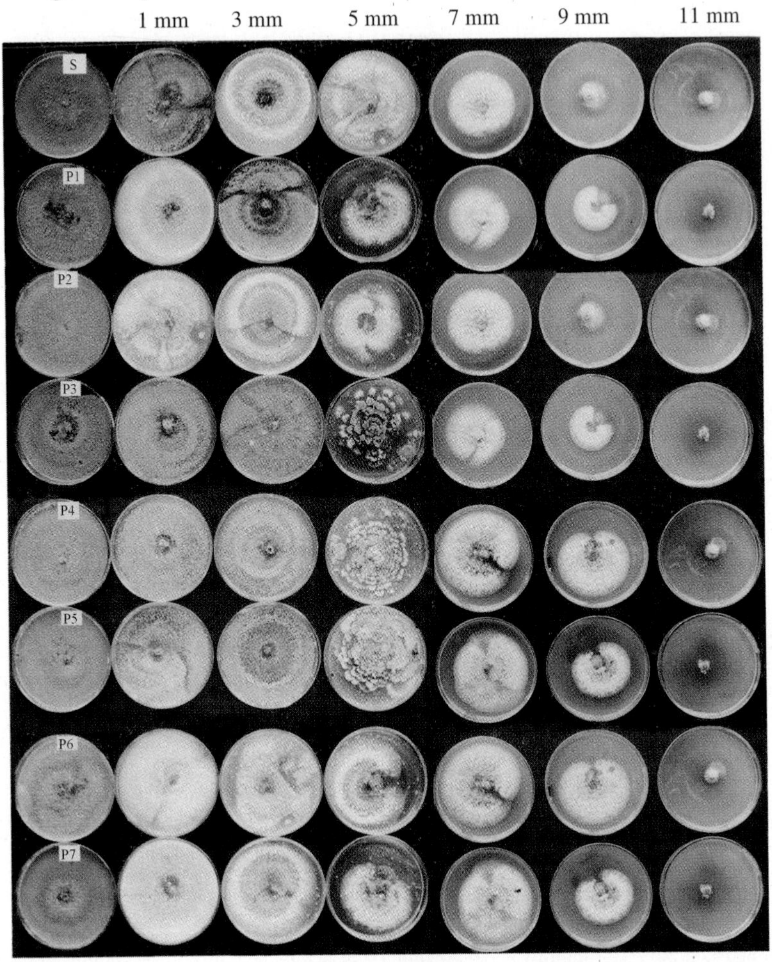

Colour Plate 3 Variation in sensitivity of regenerted protoplast isolates of *Trichoderma harzianum* to Iprodione.

S-Wild strain, P1-P14-Protoplast regenerated isolates

Source: *Karpagam, 1994*. M. Phil dissertation, University of Madras, India, 30-57

Table 23 Pathogenicity variations in regenerated protoplast cultures of *C. capsici*

C. capsici	Mean lesion diameter of 4 fruits (mm)
S	25
C1	30
C2	26
C3	25
C4	28

(*Contd.*)

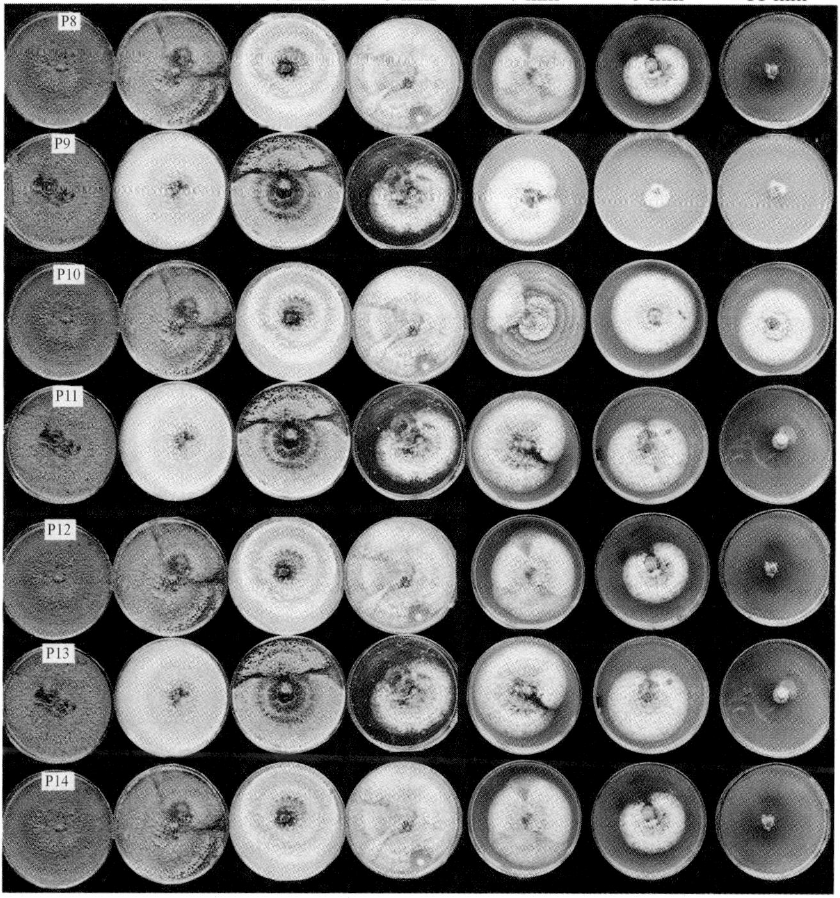

Colour plate 3 contd.

Table 3 contd.

C. capsici	Mean lesion diameter of 4 fruits (mm)
C5	28
C6	20
C7	30
C8	34
C9	30
C10	34

Source: Kalpana, 1995. M. Phil Thesis, Univ. of Madras, India 26-34C1-C10 protoplast regenerants.

Among the tested cultures C8 and C10 showed the maximum lesion diameter followed by C1, C7, C9, C4 and C5, C6, showed reduced pathogenicity than the other regenerants and source culture. Thus, most virulent and avirulent strains can be screened from protoplast regenerated strains. Protease activity is an easy test to isolate and identify a virulent strain (Table 24). The protease activity was maximum in C8 followed by C10, C4, C3 and C1. The protease activity in the C8 culture was 0.68 units/ml and it was distinctly more than the other cultures, confirming its virulence over other cultures.

Table 24 Protease activity in regenerated protoplast cultures of *C. capsici*

C. capsici	Protease activity units/ml
S	0.35
C1	0.40
C2	0.24
C3	0.41
C4	0.48
C5	0.34
C6	0.25
C7	0.32
C8	0.68
C9	0.37
C10	0.51

Source: Kalpana, 1995. M. Phil Thesis, Univ. of Madras, India, 26-34.

k. Variation in the Production of Cellulase (Kuwabara *et al.*, 1989)

Protoplasts from mycelium of *Robillarda sp.* strain Y20 on regeneration exhibited variations in the production of CM cellulase and β-glucosidase activities compared to the parents. Among the regenerants tested, 26 and 40% strains showed higher CM cellulase and β-glucosidase productivities than the parents. Regenerants from *Pleurotus ostreatus* showed an increase in fruiting body production and shorter culturing period (Magae *et al.*, 1985).

Chapter 3

Protoplast Fusion

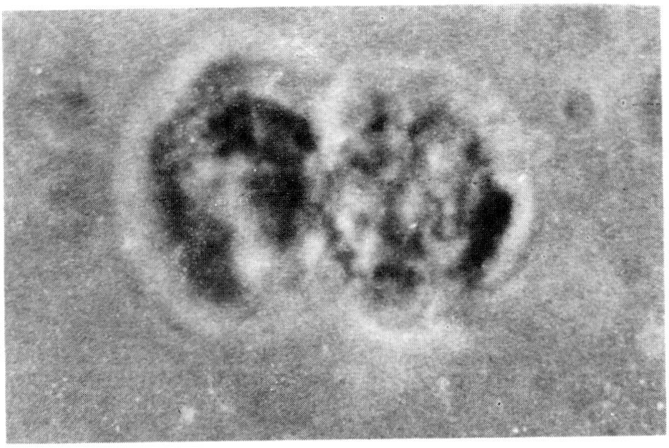

Protoplast fusion is considered to be a very powerful and promising technique for the genetic manipulation of all filamentous and yeast fungi with potential applications in industry, agriculture, pharmaceuticals and environment because it induces not only a high frequency of recombinations but also genetic interaction between different microorganisms: interspecific, intergeneric and even between different kingdoms. Usually, there is a high probability for reassortment with protoplast fusion than mycelial anastomosis, because protoplast fusion permits parental nuclei to be brought in close proximity at a much higher frequency. Success in hybrid formation has been most readily achieved when the parental strains belong to closely related fungal taxa. Protoplast fusion techniques have been used to bypass many natural barriers to cross-breeding in fungi.

Protoplast fusion has been reported in numerous species of Deuteromycetes where no sexual cycle exists. In Ascomycetes, where the sexual cycle is difficult or impossible to demonstrate, protoplast fusion has been successful in allowing genetic exchange. Protoplast fusion allows combination of entire biosynthetic pathways which are separated by species and genus barriers allowing introduction of novel characteristics into strains. In addition, it allows investigators to bypass mating type and incompatibility group barriers and to study mitochondrial genetics. Protoplast fusion is best suitable for strain improvement in fungi. Enhanced yield of mushrooms, microbial products, antagonistic strains with enhanced biocontrol potential can easily be developed.

The discovery of parasexual cycle by Pontecorvo and Roper (1952) provided a method for the formation of genetic recombinants in asexual fungi. Consequently, the parasexual cycle could be applied in addition to the classical methods of mutation and selection, for the production of potentially improved strains of industrially important fungi and biocontrol agents. The methods involved to induce heterokaryon formation were laborious and frequently inadequate. The application of parasexual cycle remained mainly restricted to formal genetic analysis of species of non-industrial importance such as *Aspergillus nidulans* or *Neurospora crassa* which readily forms anastomosis. Parasexual recombination has now been overshadowed by the development of protoplast fusion technology. Protoplast fusion is now regarded as an effective technique to study the genetics and biochemistry of fungi and yeasts and for leading to new approaches in genetic manipulation (Peberdy, 1976; 1979; 1983; Lowe and Elander, 1983; Ball, 1984b; Groves and Oliver, 1984; Kevei and Peberdy, 1984; Toyama *et al.*, 1984). With this technique, one can prepare cell-free extracts and organelles and generate novel intraspecific, interspecific, intergeneric and intrageneric hybrids in fungi, overcoming natural barriers to hybridization and opening up new possibilities in the study of gene transfer in fungi.

3.1 EXCHANGE OF GENETIC INFORMATION IN FUNGI

In both filamentous and non-filamentous fungi, there are four important methods available to exchange genetic information. 1. Sexual crosses can be used if the species has a sexual cycle and has the advantage of being very precise, but the strains must be of the same species and sexually compatible. 2. Second method is the parasexual cycle which can be used for a sexual species but this technique is again limited in that the strains must be of the same species and must be of the correct heterokaryon compatibility group. 3. A third method, transformation with recombinant DNA technology, is advantageous in that precise pieces of DNA can be moved between taxonomically diverse organisms and gene dosage can be increased readily. A disadvantage in applying this technique is that the recombinant DNA is restricted in size and the technology for protoplast formation is usually required. 4. The fourth method is the protoplast fusion and the disadvantage is that information and regeneration techniques for fungal protoplasts are not uniform to all fungal species and methods must be devised for each species and sometimes individual variant. Additionally, markers are needed in each parent strain to distinguish heterokaryons from diploids and from recombinants. The introduction of polyethylene glycol (PEG) as powerful fusogenic agent in plant protoplast systems spurred the interest in use of PEG to induce fusion in other protoplast systems.

3.2 PROTOPLAST FUSION METHODS

a. Virus-induced cell fusion

An important discovery which can be considered as the onset of induced cell fusion, was made by Okada *et al.* (1957). They first described that animal tumor cells in suspension can be rapidly fused under the influence of Hemagglutinating Virus of Japan (HVJ) or Sendai virus giving rise to the generation of multinucleate cells. Virus completely inactivated by UV retained its fusion capacity. Harris and Watkins (1965) using UV inactivated Sendai virus detected the practical application of virus-induced cell fusion to produce heterokaryons between different animal cells and they and others showed that even wide species difference was no barrier to cell fusion. A number of enveloped DNA and RNA viruses have since been reported to produce syncytia (Hoekstra and Kok, 1989), but certain RNA viruses such as those that belong to the paramyxovirus group which include mumps, New Castle disease virus and para-influenza viruses are better suited as fusion reagents. They stimulate membrane fusion during virus entry and not as a consequence of virus replication, allowing the use of inactivated virus for membrane fusion. Cell fusion by inactivated virus avoids complications such as cell death or an increased possibility of chromosome rearrangements resulting from the use of infectious virus. Instead of UV, alkylating agents including β-propiolactone can also be used for virus inactivation causing complete destruction of virus infectivity while leaving the cell fusion properties unaffected. In addition, chemically inactivated viruses fuse cells in suspension. Induction of cell fusion by viruses has been used very often in genetic experiments on mammalian cells, but has now been replaced largely by other means of fusion. However, virus induced fusions remained limited to mammalian cell types. They are not applicable, e.g. to microbial or plant protoplast fusions. Therefore, other means than virus-induced membrane fusions ought to be found to obtain membrane fusions.

b. Chemically Induced Cell Fusion

1. Sodium nitrate with Ca^{2+} ions

Power *et al.* (1970) first devised controlled conditions using high concentrations of $NaNO_3$ (0.25 M) to fuse plant protoplasts from different taxa in order to produce hybrid somatic cells. This technique did not seem efficient, had uncertain reproducibility and an extremely low fusion frequency of less than 0.01% (Lazar, 1983). It could also not be used for microbial protoplast fusions (Anne, 1977). Improved results were obtained when the cells were treated with Ca^{2+} ions under alkaline condition. Ca^{2+} ions have been implicated as modulators in many biological fusion events. Sendai-induced fusion require Ca^{2+}. The fusion capacity of Ca^{2+} solution at high pH was detected for mammalian cells by Toister and Loyter (1971). Calcium under alkaline condition led to fusion of other cell types including plant protoplasts, liverwort protoplasts, bacterial and fungal protoplasts.

Although Ca^{2+} at high pH was effective as a fusogen, the percentage of fused cells remained low.

2. Polyethylene glycol (PEG)

Improved results were obtained with a new class of fusogens, independently detected by Wallin *et al.* (1974) and Kao and Michayluk (1974) who observed that the non-ionic water soluble surfactant PEG could efficiently agglutinate plant protoplasts and that these protoplasts subsequently were fused at high frequency. It soon became clear that PEG-induced cell fusion was not cell-specific. In a short period of time, it was shown that PEG was an efficient fusogen for any kind of cells (Ferenczy, 1981) including fungal protoplasts, bacterial protoplasts and mammalian cells. It is suitable for the fusion of cells of different phylogenetic origin. Instead of PEG, the non-ionic surfactant polyvinyl alcohol (PVA) with an average polymerization degree of 500-1500 can also be used for fusion with similar efficiency as PEG (Nagata, 1978), but the latter compound is more generally applied as fusogen.

Since the discovery of fusogenic properties of PEG/Ca^{2+} by Kao and Michayluk (1974), it has been used to induce the fusion of many types of cells and organelles and continues to be most widely used method of fusion. Besides the fusion of plant, fungal and bacterial protoplasts and organelles, PEG is the fusogen of choice in somatic cell genetic studies with mammalian cells (O'Malley and Davidson, 1977; Robinson *et al.*, 1979). The fusogenic properties have also been used to fuse recombinant *Agrobacterium tumefaciens* protoplasts with plant protoplasts (Cocking, 1984) as well as artificially constructed liposomes containing DNA with bacterial, plant and fungal protoplasts (Radford *et al.*, 1981; Makins and Holt, 1981; Nagata, 1984).

The action of PEG as a fusogenic agent is not fully understood (Robinson *et al.*, 1979). PEG was first used in conjunction with protoplasts in 1953 when Weibull used a PEG solution to stabilize *Bacillus megaterium* protoplasts. Neither agglutination nor subsequent fusion was noted at that time. Other investigators have documented the necessity of small quantities (1-100 mM) of Ca^{2+} to stimulate high frequency protoplast fusion (Anne and Peberdy, 1975; Constable and Kao, 1974; Ferenczy *et al.*, 1976).

Ferenczy *et al.* (1975a) found that PEG solution is a good osmotic stabilizer if the concentration is at least 25% (w/v) and that the effect of Ca^{2+} on fusion was significant. Despite the good aggregation of protoplasts at high PEG concentrations, the fusion frequency was negligible without the addition of Ca^{2+}. The addition of as little as 1 mM $CaCl_2$ was effective in stimulating the fusion process. Prior to the discovery of the fusogenic properties of PEG/ Ca^{2+} high Ca^{2+} concentration under alkaline pH condition causes fusion (Anne and Peberdy, 1975; 1976; Keller and Melchers, 1973; Binding and Weber, 1974).

PEG may act as a polycation, inducing the formation of small aggregates (Plate 15) of protoplasts. Constable and Kao (1974), working with plant protoplasts, assumed that PEG acted as a molecular bridge between adjacent membranes, either directly by hydrogen bonds or indirectly via Ca^{2+}. Anne and Peberdy (1975) found that lower pH promotes formation of linkages via hydrogen bonds if Ca^{2+} is not present. Presence of other cations, namely K^+, Na^+, or Mg^{2+}, greatly diminishes the stimulatory effect of Ca^{2+} on fusion and attributed this effect to a decrease in the amount of Ca^{2+} bound to the membranes. They postulated that high pH might promote Ca^{2+} links with the ($-O-CH_2-$) ends of the PEG molecules. Anne (1977) had shown interesting stages of protoplast fusion of *P. chrysogenum* nutritionally complementary

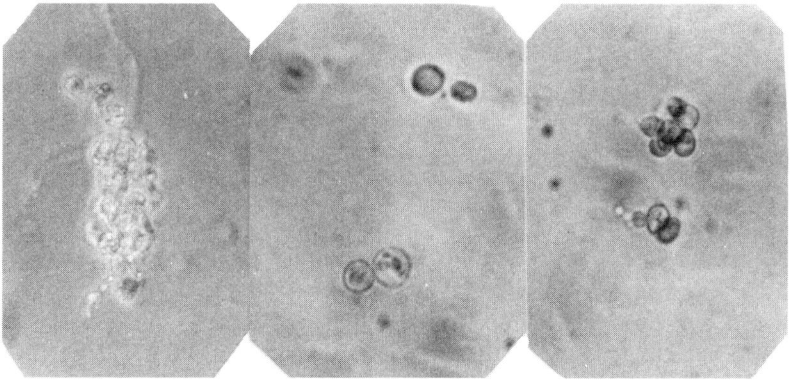

Plate 15 Small aggregates of protoplasts after PEG treatment.
Source: D. Lalithakumari, 1999 (Unpublished).

strains (Plate 15a). Protoplasts adhered firmly and shrank indicating the high osmotic pressure of the PEG solution. Fused protoplasts were observed on dilution of PEG solution.

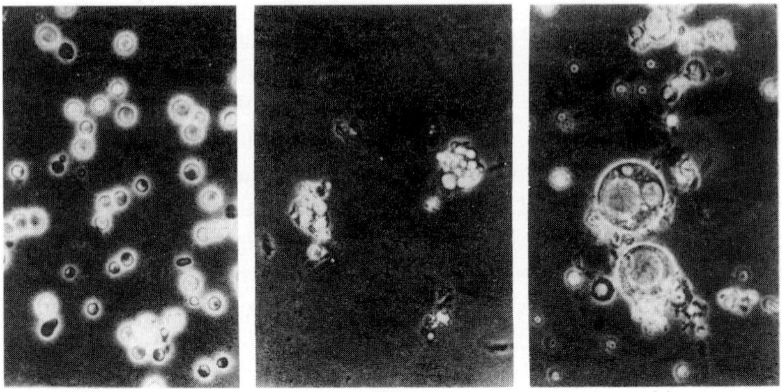

(a) (b) (c)

Plate 15a Sequential micrographs of the fusion of *P. chrysogenum lys pro/arg whi* and *leu met/cys ylo*.

(a) Protoplast suspension before PEG-treatment; (b) aggregates of shrunken protoplasts after addition of 30% (w/v) PEG in 0.01 M $CaCl_2$ and 0.05 M glycine at pH 7.5; (c) fused protoplasts after dilution of PEG with MM Reproduced from Anne, 1977.

A solution of 30% (w/v) PEG containing 10 mM $CaCl_2$ (Plate 15b) and 50 mM glycine at pH 7.5 was optimal for *V. inaequalis* Vijayapalani and Lalithakumari, 1995). The molecular weight of PEG used is critical to the fusion frequency and most protocols call for PEG 4000 or 6000 (Anne and Peberdy, 1975; Ferenczy *et al.*, 1975a). Ferenczy *et al.* (1976) studied PEG solutions with molecular weights of 400, 1540, 4000, 6000 and 20,000 for their ability to promote fusion. The best complementation frequencies were obtained with 25% (w/v) (PEG) 4000 or 6000 in 10-100 mM $CaCl_2$. In *Trichoderma*, protoplast fusion between *T. harzianum* and *T. longibrachiatum* (Plate 16) was brought about with 10 mM $CaCl_2$ + PEG 3350 (40-60%) (Mrinalini and Lalithakumari, 1993). Besides, for somatic cell fusion, PEG treatment could be used as an efficient means to bring about the uptake of macromolecules such as DNA (Hopwood, 1981) and organelles like mitochondria or nuclei (Ferenczy, 1984; Sivan *et al.*, 1990) or even whole cells into larger cell types such as bacteria into fungal protoplasts (Guerra-Tschuschke *et al.*, 1991) or into plant protoplasts (Hasegawa *et al.*, 1983) or bacteria into mammalian cells. This method is applied for direct gene transfer from bacteria to mammalian cells (Sandri-Goldin *et al.*, 1983; Caporale *et al.*, 1990). Variations in the fusogenic property could be partially due to different amounts and types of contaminants in the various PEG preparations and also due to variable optimal fusion conditions. The impurities consists of oxidative decomposition products such as aldehydes, ketones and acids which mainly cause cytotoxic effects reducing the viability of the greate cells. The impurities of PEG are also influenced by storage and autoclaving. Autoclaved PEG solution decreased yeast protoplast reversion and hybrid production Kobori *et al.* (1991) and diminished fusion of plant protoplast (Chand *et al.*, 1988). Hence, membrane sterilized PEG preparations are preferred.

3. Poly vinyl alcohols (PVA)

Polyols have been used successfully to induce protoplast fusion. Nagata (1978) used a solution containing 15% (w/w) PVA (molecular weight 500),

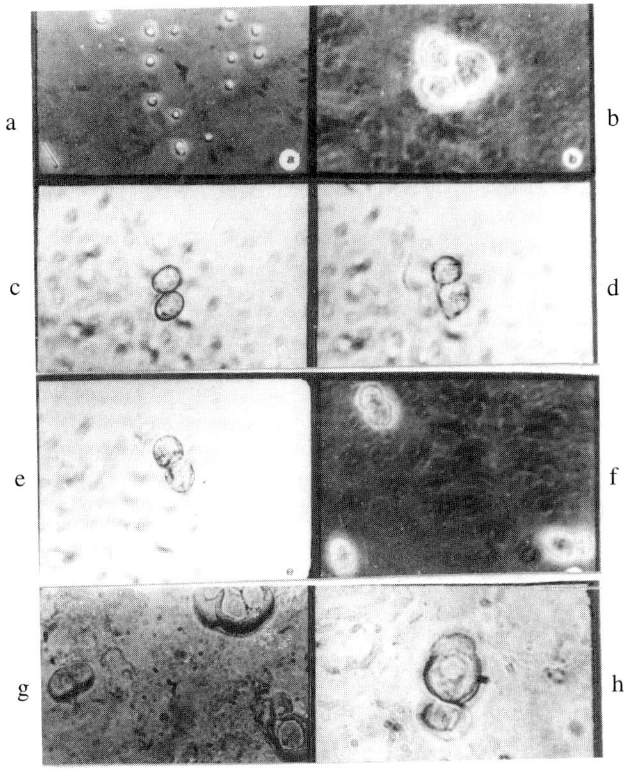

Plate 15b Fusion of protoplasts of *V. inaequalis* in PEG. (a) Free protoplasts before the addition of PEG (480×); (b) Clogging of protoplasts in 80% PEG (480×); (c) Mutual attraction of protoplasts after the addition of PEG (1,440×); (d) Adherance of protoplasts after 2 min (1,440×); (e) Adherance of protoplasts after 2 min (1,440×); (f) After 4 min of PEG treatment (1, 4409×); (g) After 6 min of PEG treatment (1,200×); (h) After 8 min of PEG treatment (1,720×); (i) Enlarged fused protoplasts (1,720×).

Source: Vijayapalani, 1995. Ph.D. Thesis, University of Madras, India, 97-225.

50 mM $CaCl_2$ and 0.3 M mannitol to induce fusion between plant protoplasts and found that PVA/Ca^{2+} induced fusion comparable to that of the PEG/Ca system, has also been used to successfully fuse artificial liposomes containing tobacco mosaic virus (TMV) RNA with protoplasts of *Nicotiana tabacum* and *Vinca rosea* (Nagata, 1984).

4. Lipids in cell fusion

Several lipids and phospholipids have been shown to induce cell fusion or to have at least a positive effect on the fusion process (Lucy *et al.*, 1971). Lysolecithin and glyceryl mono oleate (Croce *et al.*, 1971, Cramp and Lucy, 1974) have been used in mammalian cell fusion. For plant protoplasts a positively charged synthetic phospholipid has been used for fusion (Nagata *et al.*, 1979).

5. Liposomes

Liposomes also were used in combination with PEG forming lipid crystal structures. Liposomes are also synergistic fusogens in combination with PEG.

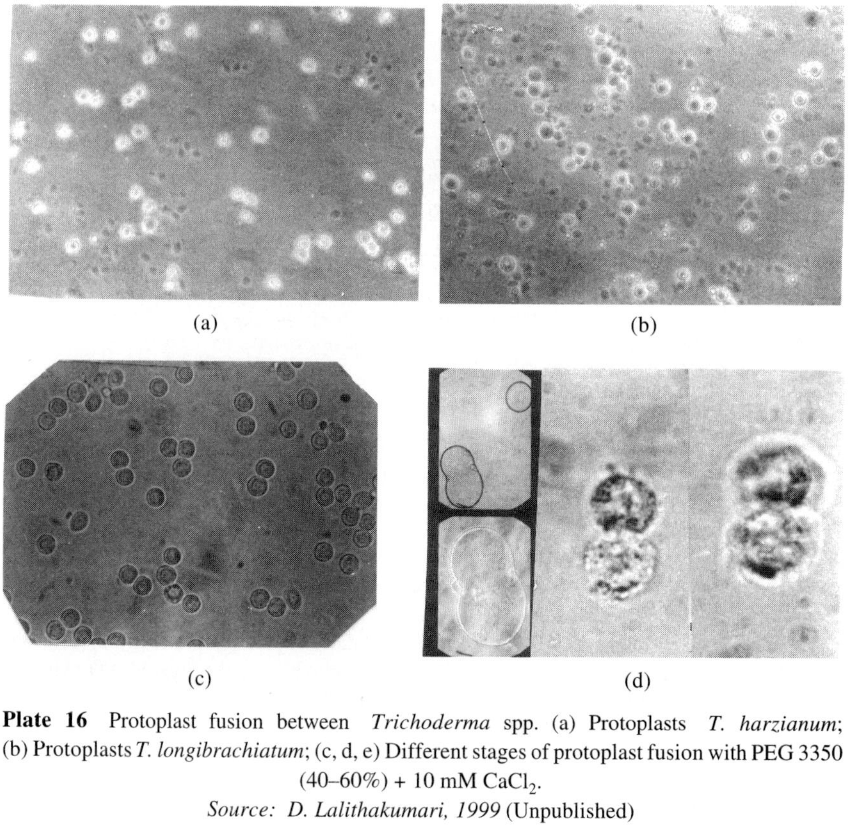

Plate 16 Protoplast fusion between *Trichoderma* spp. (a) Protoplasts *T. harzianum*; (b) Protoplasts *T. longibrachiatum*; (c, d, e) Different stages of protoplast fusion with PEG 3350 (40–60%) + 10 mM $CaCl_2$.
Source: D. Lalithakumari, 1999 (Unpublished)

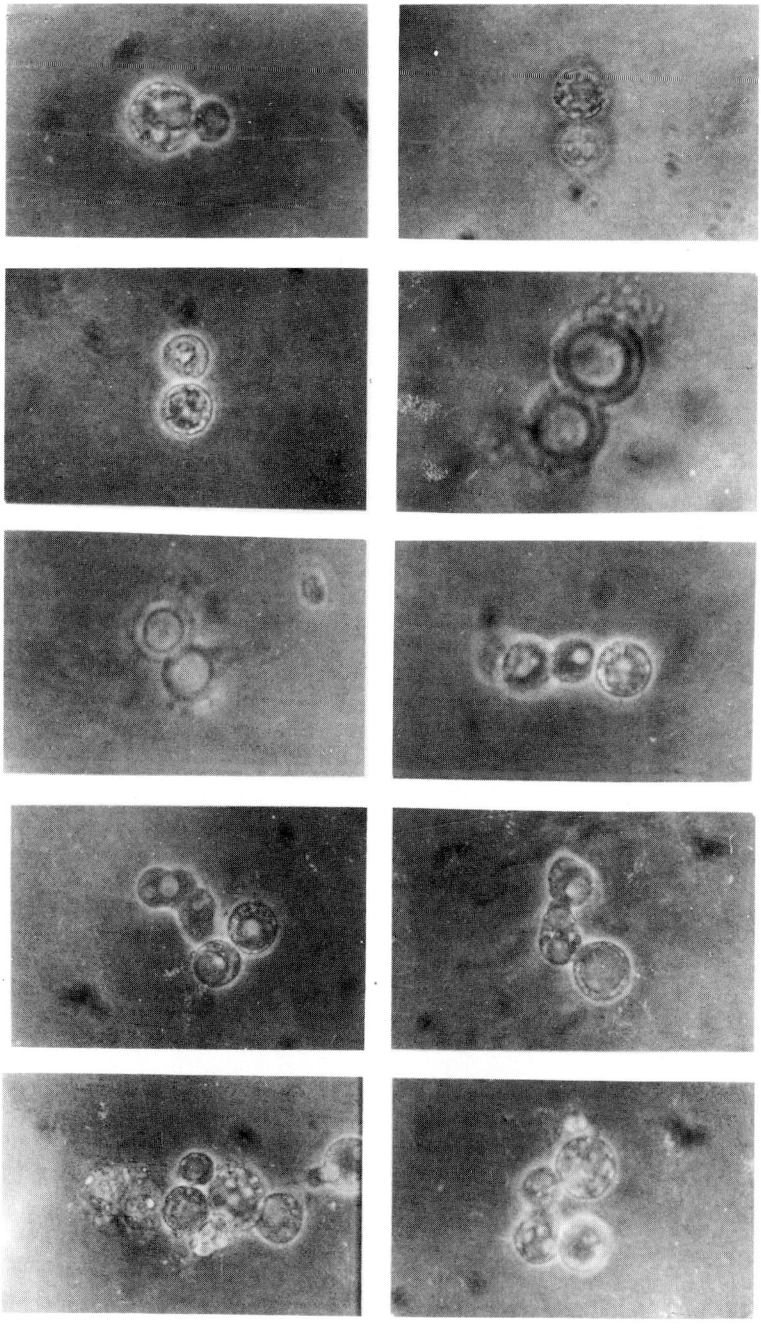

Plate 16(e)

Various stages of protoplast fusion

In the presence of liposomes, the requirement for high concentration of the polymer is lowered for efficient protoplast fusion (Makins and Holt, 1981). The major advantage of liposomes is to protect the biomolecules entrapped in the vesicles from degradation by hydrolytic enzymes. Hence, liposomes are increasingly used in transformation and transfection experiments with mammalian cells (Felgner *et al.*, 1987), bacterial protoplasts (Makins and Holt, 1981, Boizet *et al.*, 1988), and plant protoplasts (Lurquin, 1979).

c. Electric Field Mediated Protoplast Fusion

A totally new approach to *in vitro* cell fusion is the electrofusion method (Plate 17).

This method is based on the combined action of dielectrophoresis and a transient change in membrane permeability obtained by electric pulse. The existence of a transient permeability change under the influence of high electric pulses had already been suggested by Neumann and Rosenheck

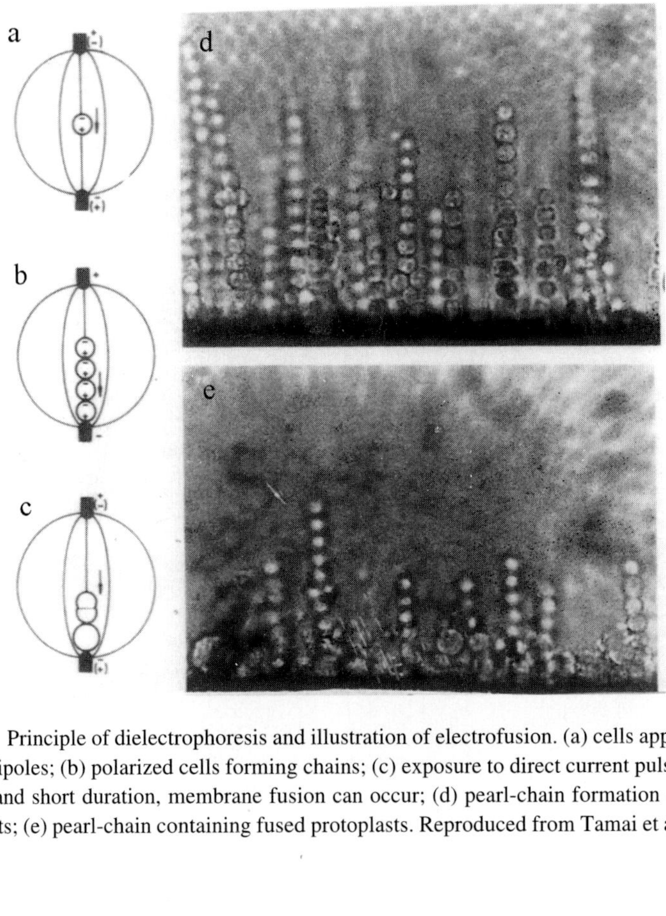

Plate 17 Principle of dielectrophoresis and illustration of electrofusion. (a) cells appearing as induced dipoles; (b) polarized cells forming chains; (c) exposure to direct current pulse of high intensity and short duration, membrane fusion can occur; (d) pearl-chain formation of fungal protoplasts; (e) pearl-chain containing fused protoplasts. Reproduced from Tamai et al., 1989.

(1972). Zimmermann (1986), Zimmermann and Urnocity (1987) investigated it more thoroughly and they exploited the permeability change to obtain electro-induced fusion. They suggested that fusion occurred as a consequence of what they called a reversible breakdown of the membrane caused by a structural alteration of the phospholipid bilayer in the membrane due to electric pulses.

Electric field induced permeabilization and fusion are now used along with PEG treatment routinely in cell biology. Electrofusion is a gentle procedure which gives high fusion frequencies under optimal conditions. Like PEG or PVA, it is applicable to all types of naked cells. Due to the inherent variation of biological systems, optimal conditions have to be determined for each cell type. It is therefore, impossible to standardize a specific method for electrofusion or electrotransfection. Compared to PEG or PVA it is claimed (Zimmermann, 1986) that electrofusion has several advantages over the chemically induced fusion including the significantly higher fusion frequency, which is certainly the case for electrotransformation and transfection and it has a less harmful effect on the cell viability. As a consequence, electrofusion and electrotransformation are gradually more frequently applied. Several types of cell fusion apparatus are now commercially available.

The breakdown voltage of most cell membranes is about 1 V, which corresponds to a field pulse in the KV/cm range. The voltage is strongly influenced by temperature. It decreases towards higher temperature, and it depends on membrane and system parameters. Membrane permeability is influenced by strength and duration of the electric pulse. Increased field strength and prolonged exposure times cause a considerable increase in membrane permeability that allows the uptake of large molecules. Very long pulse lengths at high field strength, however, give rise to an irreversible breakdown of cell membrane.

Because of increased membrane permeability, cells subjected to an experimentally determined, correct high electric field pulse can effectively be transformed or transfected. The first proof that this electrical method was effective for introduction of DNA into cells was provided by Auer *et al.*, (1976), demonstrating that DNA or RNA can be introduced into erythrocytes, after dielectric breakdown of the red blood cell membrane. Since these first experiments, the electroporation method has become a valuable and even superior alternative to the chemical transfection or transformation method for any type of cell and microbial cells can be transformed without prior protoplasting (Delorme, 1989; Chakraborty and Kapoor, 1990).

When electropermeabilized cells are brought into close contact with each other, they can be fused. The aggregation required for fusion can be obtained by dielectrophoresis. Dielectrophoresis comprises the migration of neutral particles such as cells or protoplasts in an alternating current (AC) electric

Table 25 Release, regeneration frequency and fusion frequency of protoplasts using elecroporator

	Strain	0.6 M KCl	0.6 M NH_4Cl	1.4 $MgSO_4$ & 50 mM Sod.cit.	0.6 M Suc & 0.6 M Sor
No. of protoplasts	Th	15.01	6	1.5	3
released × 10^{-6}	Tl	9.00	18	1.1	2.4
Regeneration	Th	52.9	8.2	5.32	20.5
frequency PDA (36 h)	Tl	48.6	7.9	11.3	15.6
Fusion (No PEG)		1.61	0.52	0	0.30
frequency (1% PEG)		1.70	0.50	0	0.26

Th: T. harzianum; Tl. *T. longibrachiatum*
Source: Mrinalini 1997. Ph.D. Thesis, Univ. of Madras, India 85-87.

field. In the presence of an electric field, the cell or protoplast being a neutral but highly conductive body becomes polarized and gives rise to dipole formation in which a transmembrane potential is created.The magnitude of this potential is proportional to the intensity of the external field and the diameter of the particle which has to be at least 0.3 m (Zimmermann, 1986). The positive charge of the dipole is directed towards the cathode and the negative charge is nearest to the anode. In an AC field, the field strength on both sites of the particles become unequal and the net force thus generated on the particle drives it towards the region of the higher field intensity. This is independent on the arrangement of the polarity of the electrode. This means that in an AC field the polarized but neutral particle still moves towards the region of the highest field intensity. In addition, when the particles approach, they attract each other as they are dipoles, and this leads to the formation of pearl chains (Plate 17). The number of cells within a pearl chain depends upon the population density of the cells and the distance between the electrodes, but also on the pH of the medium (Chang *et al.*, 1989). When the cells in the pearl chain are subjected to a high-intensity DC pulse they fuse as a consequence of the reversible breakdown of cell membrane.

Instead of aggregation by dielectrophoresis, electropermeabilized cells can also be fused by the addition of agglutinating agents such as PEG, PVA or spermine (Chapel *et al.*, 1986), and even simple centrifugation is sufficient (Teissie and Rols, 1986). Since agglutination may occur either before or after electroporation, it indicates that a long lived fusogenic state of the membrane can be induced by high field pulsation (Montane *et al*, 1990). The observation that this fusogenic state can last for several minutes is of particular interest from a practical point of view. Cells having large distinctive differences in their own specific field intensity for the induction of the fusogenic state can separately be electropermeabilized prior to fusion treatment.

d. Protoplast Fusion Using Electroporator

Electrofusion could be carried out in an electroporator (Hoefer Scientific Inc., USA). For fusion in the electroporator, protoplast suspension was pipetted into the fusion chamber and allowed to settle for 5 min. Fusion was initiated by application of three square wave pulses at a field strength of 125 V/cm, capacitance 100 uF and the discharge interval was 2 msec. depending on test fungi. After the application of the electric pulses, the protoplasts in the fusion chamber were checked under a microscope for completion of fusion (Table 25) Mrinalini (1997).

e. Laser Induced Cell Fusion

A very promising method that is under investigation is laser-induced cell fusion (Wiegand *et al.*, 1987). Laser beams with a very narrow diameter ranging between 0.3-0.5 μm can be produced to allow the individual fusion of selected pairs of target cells. In this manner, the time needed to select the cells after fusion can be reduced.

f. Thermo-osmotic Forces and Fluxes

Antonov (1990) suggested an alternative approach to cell fusion, based on thermodynamic factors, i.e. thermo-osmotic forces and fluxes. This method uses the induction of membrane defects by cooling the cell suspension to 0°C followed by an aggregation forced either by dielectrophoresis or temperature-controlled centrifugation. The effectiveness of this method needs intensive research. Usually, there is a high probability for reassortment with the protoplast fusion techniques compared with mycelial anastomosis because the former permits parental nuclei to be brought into close proximity, especially with uninucleate protoplasts. In hyphal cells, juxtapositioning of nuclei is infrequent (Hamlyn and Ball, 1979). Success in hybrid formation has been most readily achieved when the parental strains belonging to closely related fungal taxa such as *A. nidulans* + *A. rugulosus* (Kevei and Peberdy, 1979), *P. chrysogenum* + *P. cyaneofulvum* (Peberdy *et al.*, 1977) or *P. cyaneofulvum* + *P. citrinum* (Anne and Eyssen, 1978). In all the crosses the hybird had a phenotype totally different from the parental strains and exhibited hetero-zygosity for all the genetic markers used. In some instances, it was shown that the hybrid phenotype was a consequence of nuclear fusion (Kevei and Peberdy, 1979, 1984). Instability in the hybrids, either spontaneous or induced, resulted in the segregation of progeny with novel phenotypes, which in *A. nidulans* + *A. rugulosus* hybrid was shown to be a consequence of random chromosome assortment (Kevei and Peberdy, 1979).

3.3 STRATEGIES FOR SELECTING FUSION PRODUCTS

a. Auxotrophic Mutants

Because of the enormous problem in identifying true fusion products when wild-type parent strains are used as fusion partners, some type of selective system or medium must be employed to ensure that only fusion products regenerate into colonies. The most common method to date has been, the use of complementary auxotrophic mutants as fusion partners. The fusion mixture is plated on a minimal medium, one on which neither parent strain can grow but on which the fusion products can grow.

Genetically marked strains such as auxotrophs having specific nutritional requirements or strains with mutant spore colour, are essential for the detection of heterokaryons, diploids and recombinants. For example, the auxotrophic strains which show nutritional complementation upon cytoplasmic fusion can be selected on a minimal medium as prototrophic heterokaryons.

Spore color markers, besides being helpful in heterokaryon selection, are also useful for the identification of diploids formed after nuclear fusion in the heterokaryons (Anne, 1977).

b. Drug Resistance and Antibiotic Markers

Another common method for selection of fusion products is the use of drug resistance and antibiotic markers. This method allows at least one fusion partner to be a wild-type organism, provided the wild-type strain is sensitive to the antibiotic or drug in question. The other member of the fusion pair is usually auxotrophic and carries the elements responsible for conferring resistance to the particular antibiotic or drug. The medium for regeneration of fused protoplasts consists of a minimal medium to which sufficient antibiotic or drug has been added to prevent the growth of the sensitive, wild-type fusion partner. The auxotrophic fusion partner cannot grow since it is plated on a nutritionally deficient medium. It should be noted that not all alleles for drug resistance are dominant. This method has been used extensively in yeast species.

Antibiotic resistant mutants could easily be developed in most industrially important yeast strains by either spontaneous mutation or induction with Mn^{2+}. Chloramphenicol and erythromycin resistant mitochondrial mutants of *Saccharomyces diasticus* and erythromycin resistant mutants of *S. cerevisiae* were used as markers for fusion. Rho-mutants of *S. cerevisiae* was obtained by acriflavin treatment. Respiratory competent (Rho^+) antibiotic resistant fusion products were the end result. Rho^+ cells are easily distinguished from Rho^- by their ability to ferment glycerol. In *S. cerevisiae*, resistance to chloramphenicol was obtained.

c. Fungicide Resistance Markers

Fungicide resistance markers are mostly recommended for rDNA technology to screen the recombinants. Since fungicide resistance in many cases is stable and constitutive for 10 generations it is used as marker to screen protoplast fusants.

Fungicide resistance mutants could easily be developed through mutagenesis. Fungicide resistance markers could also be developed through transformation of cloned gene incurring fungicide resistance. *Neurospora β*-tubulin resistance to benzimidazole is cloned in *E. coli* plasmid, this could be transferred into sensitive fungal protoplasts and can be used as marker for screening protoplast fusants. Mrinalini (1997) and Vijayapalani (1995) used penconazole resistance and carbendazim resistance as markers to screen penconazole + carbendazim resistant fusants of *V. inaequalis*. In a fusion between *T. harzianum* and *T. longibrachiatum* carbendazim and copper sulphate resistance was used as markers (Fig. 16) to screen the fusants (Mrinalini, 1997).

Resistance to fungicides developed due to mutation in many phytopathogens has been reported by many workers and the fungicide resistance can be successfully used as marker in protoplast fusion.

d. Inactivated Protoplasts and Dead Donors

Another selection strategy which was originally developed for use in bacterial protoplast fusion is the use of inactivated protoplasts as one of the fusion partners. One of the fusion partners can be inactivated or killed by use of brief heat treatment (Peberdy, 1989; Ferenczy, 1984) or by exposure to ultraviolet light (Hopwood, 1981), theoretically allowing the utilization of a wild-type organism as a fusion partner. Ferenczy (1984) recently applied this technique to protoplasts of *A. nidulans*. One strain had the genotype ade-(adenine), PABA-(para amino benzoic acid) and the other fusion partner was lys-(lysine), PABA-. Among the fusion products, a high proportion exhibited an altered morphology (typical of diploids or aneuploids in contrast to heterokaryons) if one of the fusion partners was heat-treated. If neither partner was heat-treated, more than 99% of the fusion products were heterokaryons and less than 1% were diploids. On the other hand, if the ade-partner was heat-treated for 6 min at 50°C, approximately 80% of the fusion products were heterokaryons, 10% were diploids, and 10% were aneuploids. This technique not only adds selective pressure but also allows isolation of a high proportion of desirable fusion products, namely hybrids of diploid and aneuploid nature *vs* almost exclusively heterokaryons.

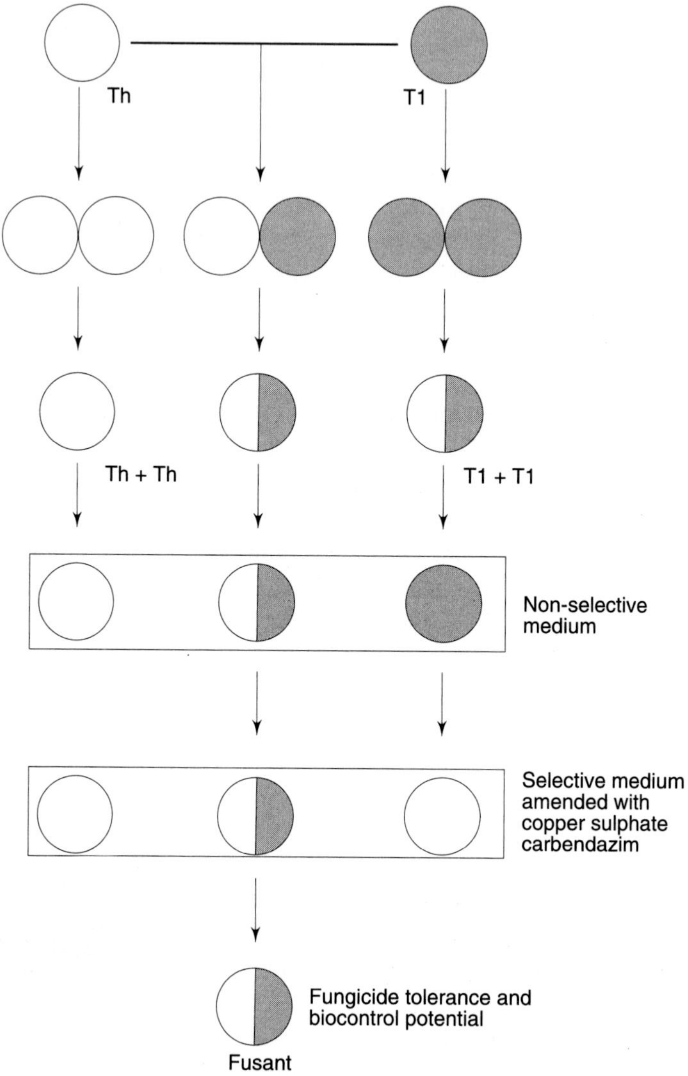

Fig. 16 Schematic diagram for selection of fusants using fungicide resistance as markers.

e. Spore Color Marker

Spore color marker has been used in protoplast fusion of both *Aspergillus* (Table 26), (Dales and Croft, 1997; Kevei and Peberdy 1977; Bradshaw *et al.*, 1983; Das *et al.,* 1989) and *Penicillium* (Peberdy *et al.,* 1977; Anne and Eyssen, 1978; Anne and Peberdy, 1976) species to aid in the detection of fusion products.

Table 26 Segregation of spore colour markers in cross 8-11 in *Aspergillus* sp.

		Albino	Olive
arg	+	(8)	90
	–	16	(4)
Met	+	11	(49)
		(13)	95
nic	+	(15)	95
	–	13	(-)
his	+	18	(79)
	–	(6)	(18

Number of recombinant genotypes shown in parentheses.
(Reproduced from Das *et al.*, 1989). Enzyme Microbe. Techno 11; 2-5

This detection system, however, has been used secondary to either auxotrophic and/or drug-resistant markers. Recently, spore colour markers were used as the only means of selection for protoplast fusion products of *T. reesei* QM 9414, a production strain used for conversion of lignocellulosic waste to soluble sugars (Ogawa *et al.*, 1987). Wild-type *T. reesei* conidiospores are green. A white-spored mutant, isolated spontaneously and a brown-spored mutant, obtained following treatment with UV irradiation, were used as fusion products. Fusion products were fawn coloured and segregated spontaneously to form either white or brown conidia. If the heterokaryon was plated on a medium containing d-camphor, diploids were formed which produced the wild-type green conidia.

f. Spore or Conidiophore Size

Spore size of conidial diameter has been used to classify fusion products as hybrids. The technique has been used most extensively in interspecific fusants of *Penicillium* (Anne and Eyssn, 1978). Tentative diploid colonies were confirmed on the basis of prototrophy, color of spores, spore size besides their doubled DNA content. Spore size was considered by Mrinalini (1997) to select fusants of *Trichoderma* spp.

g. Colony Morphology

Colony morphology has been used to identify interspecific and intergeneric fusion products especially if the species differ greatly in colony morphology. The fusants of *T. harzianum* and *T. longibrachiatum* exhibited mixed morphology acquired from both the parent strains (Mrinalini, 1997).

h. Use of Irreversible Biochemical Inhibitors

Use of irreversible biochemical inhibitors has also been used as a strategy for selection of fusion products. Wright (1978) first introduced this technique, using diethylpyrocarbonate and iodoacetamide as irreversible biochemical inhibitors. Using this strategy obviates the need for the time consuming and tedious process of mutation and identification of auxotrophs. Both fusion partners are treated with a lethal dose of inhibitor and then fused. Unfused parental cells and homokaryons do not have their damaged molecules replaced and die. Only heterokaryons receive the full complement of molecules needed for cell survival. The technique has been used successfully to produce viable heterokaryons in animal cells (Wright, 1978) and plant protoplasts (Nehls, 1978). Peberdy (1980) attempted this strategy using a variety of fungal protoplasts as fusion partners and found that the concentration and exposure times used in the animal cell and plant protoplast experiments proved inadequate with fungi.

i. Isolated Nuclei

A selection strategy which has been successfully applied in *S. cerevisiae* fusion is the use of isolated nuclei as one of the fusion partners. Ferenczy and Pesti (1982) isolated nuclei of an adenine requiring respiratory deficient mutant of *S. cerevisiae* and fused them with protoplasts of a strain auxotrophic for histidine, leucine and uracil. Both strains were of the same mating type. The resulting fusion products wee nutritionally complemented and genetically consisted of heterokaryons and stable diploids. The nuclei were isolated by lysing the protoplasts of the donor strain and then purified in a discontinuous sucrose gradient.

j. Mitochondria and Mitochondrial Transfer

Another selection strategy which has been used extensively in both intra and interspecific fusions in yeast species is the use of mitochondrial transfer or mitochondria-protoplast fusion. Selection of fusion products is based on complementation of auxotrophic markers and mitochondrially inherited drug resistance markers (Maraz & Subik, 1981; Gunge & Sakaguchi, 1979; Yoshida, 1979; Yoshida & Takeuchi, 1980; Earl *et al.*, 1981; Sakaguchi *et al.*, 1980) or restoration of the respiratory-sufficient characteristic, using a petite or mitochondria less strain as one of the fusion partners or both (Maraz & Subik, 1981; Ferenczy & Maraz, 1977; Allmark *et al.*, 1978; Morgan *et al.*, 1977; Gunge & Sakaguchi, 1979; Yoshida, 1979; Yoshida & Takeuchi, 1980.

3.4 INFLUENCE OF DIFFERENT PARAMETERS ON THE FUSION FREQUENCY

a. Cations

Cations present in the PEG solution affected protoplast fusion, but had no influence on their viability except if 0.1 M calcium or magnesium ions were present at high pH. Ca^{2+} as $CaCl_2$,0.01 M promoted the highest level of fusion at pH 7.5 (Fig. 17). But with increasing pH the optimum molarity for fusion increased to 0.6 M $CaCl_2$ at pH 9.0. Mg^{2+} as $MgSO_4$ was stimulating to a lesser degree, but at 0.1 M it was inhibitory. Increasing concentrations of Na^+ or K^+ either as chloride or as nitrate reduced fusion to a minimum and with 0.005 M EDTA it was almost completely inhibited. Salt concentrations at 0.0001 M had no influence and gave the same fusion frequency as PEG solutions, without any salt addition buffered at pH 7.5 with 0.05 M glycine and 0.01 M $CaCl_2$ with other salts at different concentrations (0.0001 M) resulted in a decrease of the degree of fusion. Among the various concentrations of $CaCl_2$, 10mM was the optimal concentration and fusion frequency decreased with increasing concentrations of $CaCl_2$ (Table 27) in *V. inaequalis* (Vijayapalani, 1995). Fusion was absolutely arrested at 100 mM

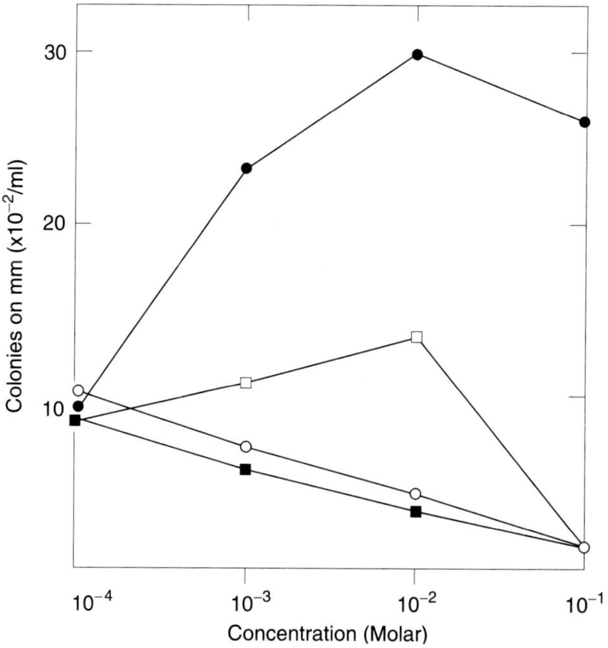

Fig. 17 Concentration (Molar). Influence of different ions on the yield of balanced heterokaryons of *P. chrysogenum* developing on MM after PEG treatment. Protoplasts (5.5×10^6 of each auxotroph) were treated (30°C, 10 min) with 1 ml of a solution of 40% (w/v) PEG in 0.05 M glycine, pH 7.5 with 0.01 M NaOH. $CaCl_2$ (–●–) $MgSO_4$ (–□–) $NaNO_3$ (–■–) or KCL (–o–) were added at various concentrations. The average number of protoplasts reverted on PM was 4.3×10^3.

Reproduced from Anne, 1977.

b. pH

Fusion occurred in PEG solutions at pH levels lower than pH 7.0. However if $CaCl_2$ was present, alkaline conditions had an important influence on fusion frequency (Ff), because the largest yield, of fusion was obtained at pH 9.0 (Fig. 18). Above pH 9.0 PEG solutions were harmful causing reduced viability in the protoplasts, although they could survive in hypertonic solutions without PEG, Ca^{2+} or Mg^{2+} at pH 10.5. In *V. inaequalis* the fusion frequency was high at pH 7.5 (Vijayapalani, 1995). Mrinalini (1997) has also reported

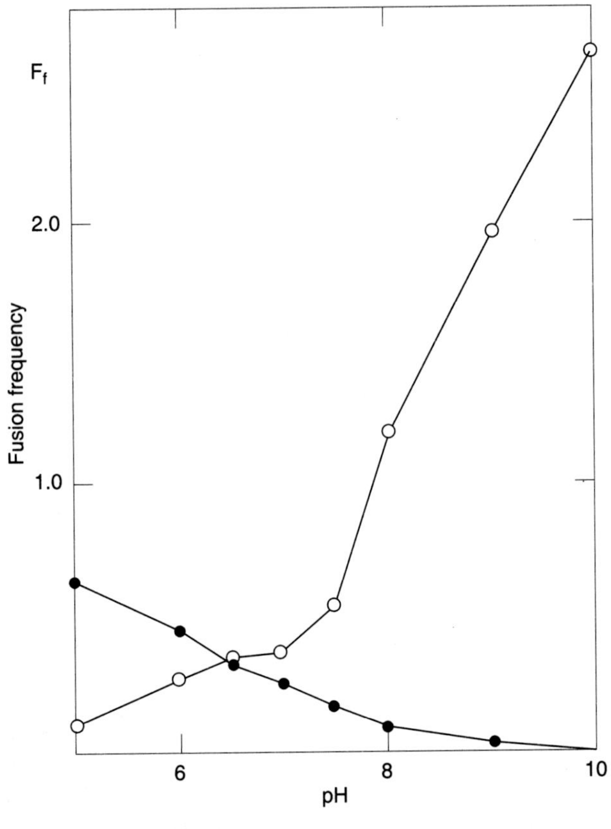

Fig. 18 Influence of pH on the fusion frequency of *P. chrysogenum* auxotrophic protoplasts in the absence or presence of Ca^{2+}. Protoplasts (0.95×10^6 of each auxotroph) were incubated (30°C, 10 min) in 40% (w/v) PEG solutions containing 0.05 M $CaCl_2$ (–o–), or without $CaCl_2$ (–•–). pH was adjusted with 0.05 M glycine or NaOH. After dilution and washes in 0.7 M NaCl, protoplasts were plated on to hypertonic MM and PM to count the number of heterokaryons and the total number of viable protoplasts, respectively, and to estimate the fusion frequency. Reproduced from Anne, 1977.

Table 27 Influence of PEG concentration on the fusion frequency of protoplast in *Trichoderma* spp.

Conc. of PEG (%)	Osmotica	Fusion Frequency
30	0.6 M KCl	0.45
	0.6 M NH_4Cl	0.42
	0.6 M Suc. &	0.31
	0.6 M Sor.	0.29
	1.4 M $MgSO_4$ &	
	50 mM Sod. Cit	
40	0.6 M KCl	1.69
	0.6 M NH_4Cl	0.543
	0.6 M Suc. &	0.68
	0.6 M Sor.	0.488
	1.4 M $MgSO_4$ &	
	50 mM Sod. cit	
50	0.6 M KCl	0.54
	0.6 M NH_4Cl	0.36
	0.6 M Suc. &	0.45
	0.6 M Sor.	0.336
	1.4 M $MgSO_4$ &	
	50 mM Sod. cit	
60	0.6 M KCl	0.568
	0.6 M NH_4Cl	0.68
	0.6 M Suc. &	0.41
	0.6 M Sor.	0.525
	1.4 M $MgSO_4$ &	
	50 mM Sod. cit	

Source: Mrinalini, 1997. Ph. D. Thesis, University of Madras, India 85-86.

7.5 pH (Fig. 19) as optimum for fusion between *T. harzianum* and *T. longibrachiatum* protoplasts.

c. Influence of Concentration of PEG

The concentration of PEG was not critical within limits (Fig. 20). Protoplast aggregates composed of up to 20 protoplasts were observed after addition of PEG solutions, if PEG concentrations used were at 20% (w/v) or higher.

However, concentrations did not stabilize and protoplasts burst. Solutions containing 30% PEG were optimal and they stabilized with the smallest reduction in the number of protoplasts. PEG at 40% or higher was very hypertonic and the protoplasts shrunk, but after dilution with minimal medium (MM) they regained their normal size. After removal of PEG by 0.7 M NaCl, the aggregates separated and larger protoplasts were observed. They were assumed to originate from fused protoplasts. If 10% PEG solutions osmotically balanced with 0.5 M mannitol and 0.4 M NaCl were used at least

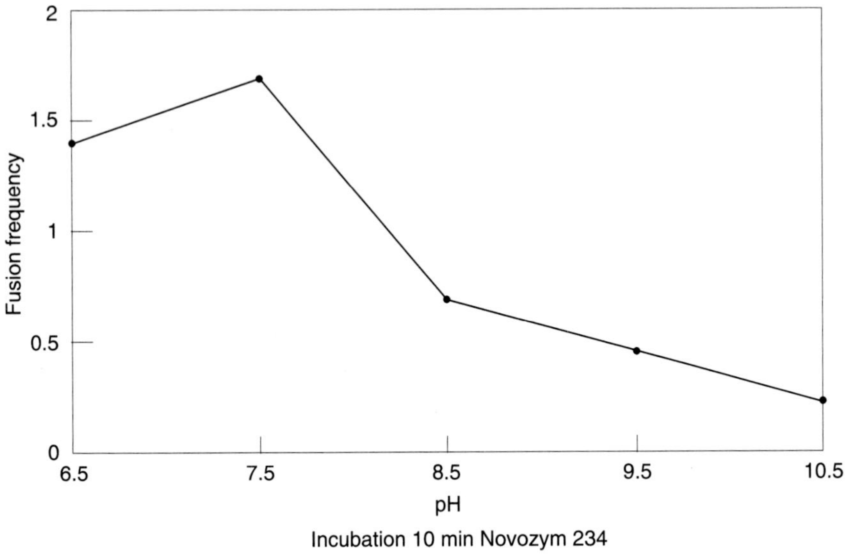

Fig. 19 Influence of pH on the fusion frequency.
Source: Mrinalini, 1977. Ph.D. Thesis, University of Madras, India, 85-86.

0.01 M $CaCl_2$ was needed for fusion and Ff was more pH dependent. If no PEG was added to the fusion inducing solution fusion was only achieved if 0.005 M Ca^{2+} were present at high pH or from pH 7.5 in solution containing 0.6 M $CaCl_2$. The effect of PEG concentration on the frequency of prototrophic colonies of *Rhizoctonia solani* was shown by Hashiba and Yamada (1984). High frequency of protoplasts colonies were obtained with 40-60% PEG. Incubation of protoplasts with PEG at concentrations 20% caused protoplast lysis.

Table 27 clearly shows that 40% and 60% (w/v) PEG were effective in bringing about high fusion frequency between. *T. harzianum* and *T. longibrachiatum* (Mrinalini, 1997). However concentrations of PEG did not stabilize the protoplasts and therefore resulted in bursting of protoplasts. At 80% PEG, clogging of protoplasts was frequently observed in *V. inaequalis* (Table 28). Only 60% PEG was optimal for the fusion of protoplasts of *V. inaequalis*.

d. Influence of Temperature on the Fusion Frequency (Ff)

Temperature plays a distinct role on Ff. Fusion of protoplasts took place at 4°C but increased with increasing temperature (Fig. 21). The number of regenerating protoplasts steadily decreased and dramatically fell when incubated above 37°C. Higher temperature had probably a dual effect

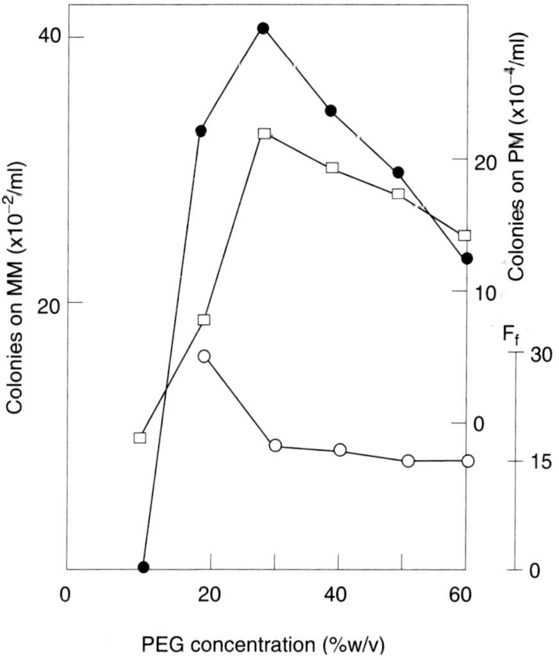

Fig. 20 PEG Concentration (% w/v). Influence of concentration of PEG on the fusion frequency (–o–) of *P. chrysogenum lys whi* and *leu met/cys ylo* protoplasts. Protoplasts (2×10^5 of each auxotroph) were treated with different concentrations of PEG in 0.01 M of $CaCl_2$ and 0.05 M glycine, pH 7.5 (30°C, 10 min). Heterokaryons and total number of viable protoplasts were counted on MM (–o–) and on PM (–•–), respectively, after 72 h.
Reproduced from Anne, 1977.

decreasing the viscosity of the PEG solution facilitating contact with protoplasts and making the cytoplasmic membranes more fusogenic by an increase in membrane fluidity (Ahkong *et al*., 1973). From Fig. 22 it is obvious that of all the temperatures used in incubation, the best fusion frequency took place at 28°C between *T. harzianum* and *T. longibrachiatum* protoplasts (Mrinalini, 1996). In *V. inaequalis* among the various temperatures tried (Table 28) 36°C was optimal for maximum fusion frequency.

e. Influence of Exposure Time of PEG

The effect on aggregation and fusion occurred immediately after PEG was assumed to have spread over the surface of the protoplasts. The increase of fusion (Fig. 23) during the first minute was supposed to be due to the time needed for entirely coating of the protoplasts by PEG or the time needed for the reaction of all fusogenic factors in the incubation mixture with the

Table 28 Effect of various factors on protoplast fusion in *V. inaequalis*

PEG (%)	$CaCl_2$ (mM)	pH	Temp. (°C)	Time (min)	Fusion Freq. (%)
20	10	7.5	36	10	—
40	10	7.5	36	10	1.7
60	10	7.5	36	10	3.2
80	10	7.5	36	10	—
100	10	7.5	36	10	—
60	25	7.5	36	10	1.3
60	50	7.5	36	10	0.8
60	100	7.5	36	10	—
60	10	6.5	36	10	—
60	10	7	36	10	3.0
60	10	8.0	36	10	—
60	10	8.5	36	10	—
60	10	7.5	30	10	1.3
60	10	7.5	32	10	2.0
60	10	7.5	34	10	2.9
60	10	7.5	38	10	3.0
60	10	7.5	40	10	2.1
60	10	7.5	36	15	0.9
60	10	7.5	36	20	—
60	10	7.5	36	25	—
60	10	7.5	36	30	—

Source: Vijayapalni, 1995. Ph.D. Thesis, University of Madras, India 217-219.

protoplast surface. Longer incubation does not increase the Ff. PEG treatment for 10 min (Table 28) favoured fusion frequency in *V. inaequalis.* Increased time of exposure leads to loss in viability of the protoplasts (Vijayapalani, 1995). Of the different exposure times to PEG used (Table 29) 10 min of exposure was optimal for maximum fusion frequency between *T. harzianum* and *T. longibrachiatum* (Mrinalini, 1997).

f. Effect of Storage of Protoplasts on Fusion Frequency

Storage of protoplasts at 4°C either in 0.7 M NaCl or in the lytic enzyme solution improved Ff (Fig. 24) though there was a loss of viable protoplasts. In lytic enzyme solution Ff continued raising but in 0.7 M NaCl fusion yield decreased faster than protoplasts reduction after 20 h storage, though the released protoplasts could be cryopreserved for a period of two to three months. Fusion frequency is maximum (Fig. 25) only when fusion takes place within 24 h of release. After 24 h, the fusion frequency declines (Mrinalini, 1997).

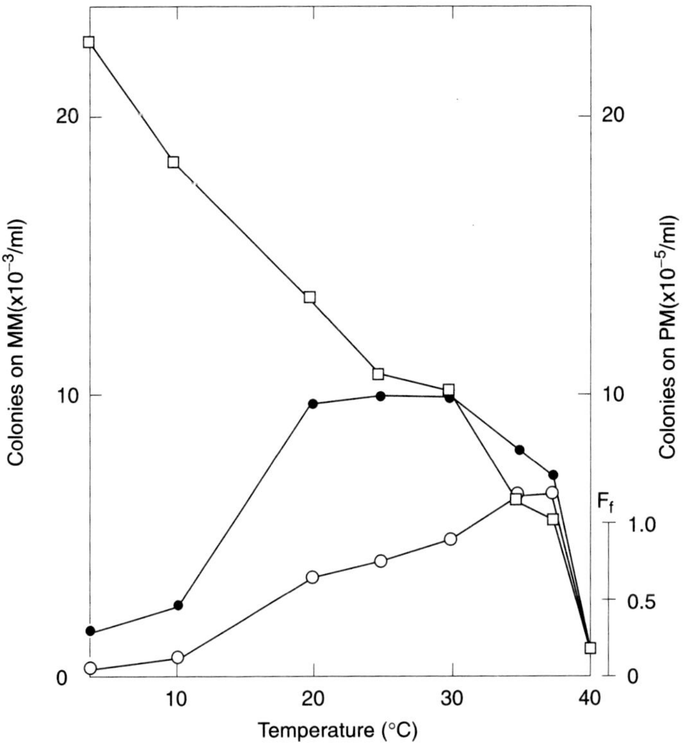

Fig. 21 Influence of temperature on fusion frequency (–o–) of *P. chrysogenum* protoplasts. One ml of 40% (w/v) PEG, pre-cooled or pre-warmed to the proper temperature, was added to 7.4×10^6 protoplasts of each auxotrophs and incubated for 10 min at different temperatures. PEG was diluted with MM, pre-cooled or pre-warmed, and protoplasts were washed with 0.7 M NaCl at the corresponding temperatures. Heterokaryons and total number of viable protoplasts were counted on MM (–•–) and on PM (–□–), respectively, after 72 h. Reproduced from Anne, 1977.

3.5 MECHANISM OF MEMBRANE FUSION

Every cellular membrane carries out highly specialized functions. Although every membrane has the same basic phospholipid bilayer structures, a different set of membrane proteins enable each cellular membrane to carry out its distinctive activities. The plasma membrane that surrounds every cell, for example, contains a set of specific proteins that affect many aspects of the cell's behaviour. Certain proteins provide anchors for cytoskeletal fibers or for components of the extracellular matrix that gives the cell its shape. Still others bind signalling molecules, provide a passageway across the membrane for certain molecules or regulate the fusion of the membrane with other

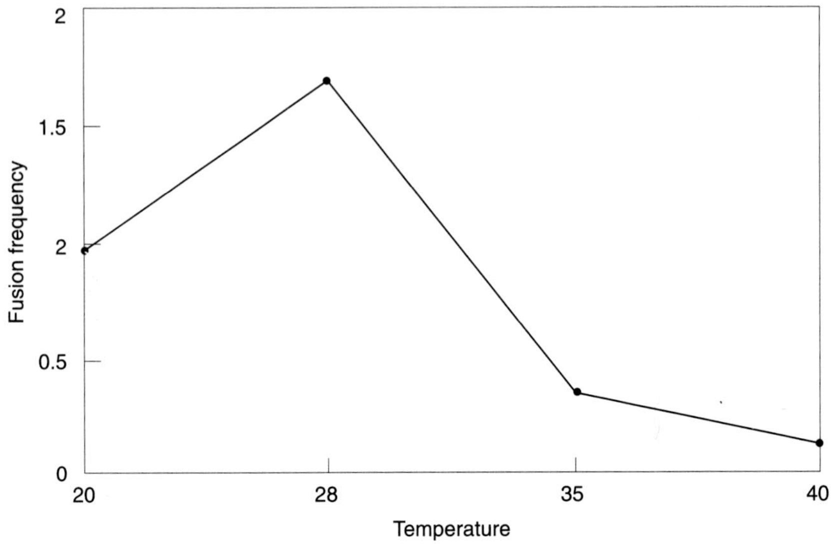

Fig. 22 Influence of temperature on the fusion frequency of *Trichoderma* spp.
Source: Mrinalini, 1997. Ph.D. Thesis, University of Madras, India, 24-26.

Table 29 Influence of exposure time to PEG on the fusion frequency

Time (min)	Conc. of PEG (%)	Fusion frequency
5	30	0.33 q 0.020
	40	0.41 q 0.005
	50	0.31 q 0.001
	60	0.43 q 0.000
10	30	0.45 q 0.052
	40	1.62 q 0.040
	50	0.54 q 0.011
	60	1.41 q 0.010
15	30	0.23 q 0.010
	40	0.31 q 0.001
	50	0.20 q 0.000
	60	0
20	30	0
	40	0.21 q 0.000
	50	0
	60	0

Source: Mrinalini, 1997. Ph.D. Thesis, University of Madras, India 32-34.

membranes in the cell. A multitude of internal membranes in each eukaryotic cell enclose separate compartments that perform specialized tasks.

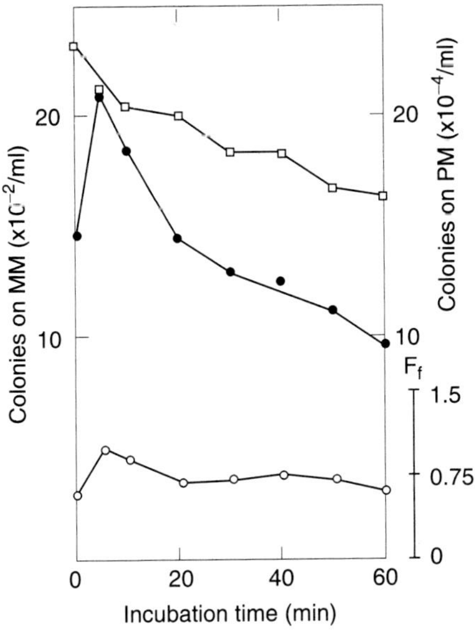

Fig. 23 Influence of exposure of PEG on fusion frequency (–o–) of *P. chrysogenum* Protoplasts. One ml of 40% (w/v) PEG in 0.01 M $CaCl_2$ and 0.05 M glycine, adjusted to pH 7.5 with NaOH, was added to a mixture of 2.7×10^6 protoplasts of each auxotrophs and incubated at 30°C for different times. After dilution with MM, and washe with 0.7 M NaCl, Fusion frequency was estimated by the number of heterokaryons developed on MM (–•–) after 72 h and by the total number of viable protoplasts counted on PM (–□–).
Reproduced from Anne, 1977.

All membrane phospholipids are amphipathic, having both hydrophilic and hydrophobic portions. By synthesizing a diverse array of phospholipids, as well as steroids such as cholesterol, cells maintain an appropriate fluidity of their plasma membranes as well as of all internal membranes.

Phospholipid bilayers have several important features including an enormous flexibility and the property of self-closing. In a phospholipid bilayer, the molecules can freely move and bilayers tolerate all kinds of deformation without disrupting the bilayer structure. As a result of this flexibility, biological membranes can be fused either spontaneously or after induction by external means. Spontaneous fusion events outside biologically controlled processes have been mentioned for all types of cells (fungal and plant protoplasts) but the frequency of fusion was very low in each case, even if fusion was forced by mechanical pressure (Ferenczy, 1981), and it remained restricted to cells of the same origin. In consequence, spontaneous fusion events were not really applicable to genetic studies. Therefore, other methods have to be devised to increase the fusion frequency.

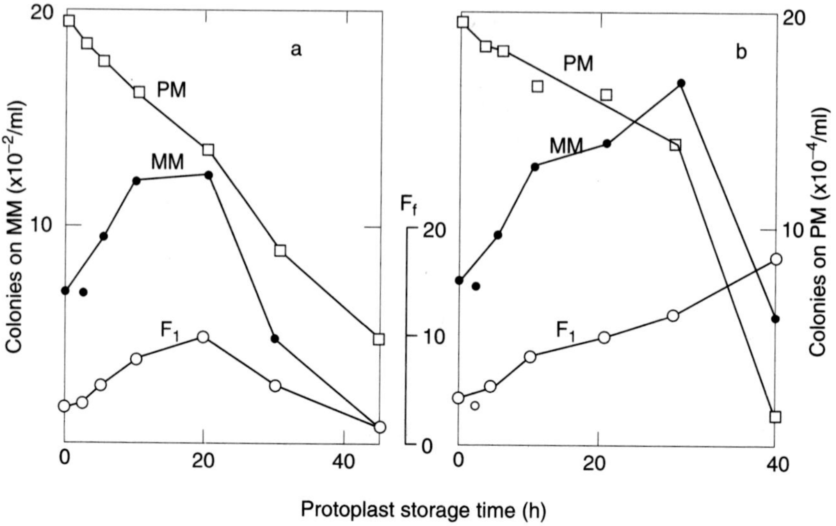

Fig. 24 Protoplast storage time (h). Effect of storage at 4°C. (a) in 0.7 M NaCl and (b) in lytic enzyme solution, on the fusion frequency (–o–) of *P. chrysogenum* protoplasts. Protoplasts (2.1×10^6 of each auxotroph) were treated with 40% (w/v) PEG in 0.01 M $CaCl_2$ Plus 0.05 M glycine, pH 7.5 (30°C, 10 min). Heterokaryons and total number of viable protoplasts were counted on MM (–●–) and on PM (–□–), respectively after 72 h.

Reproduced from Anne, 1977.

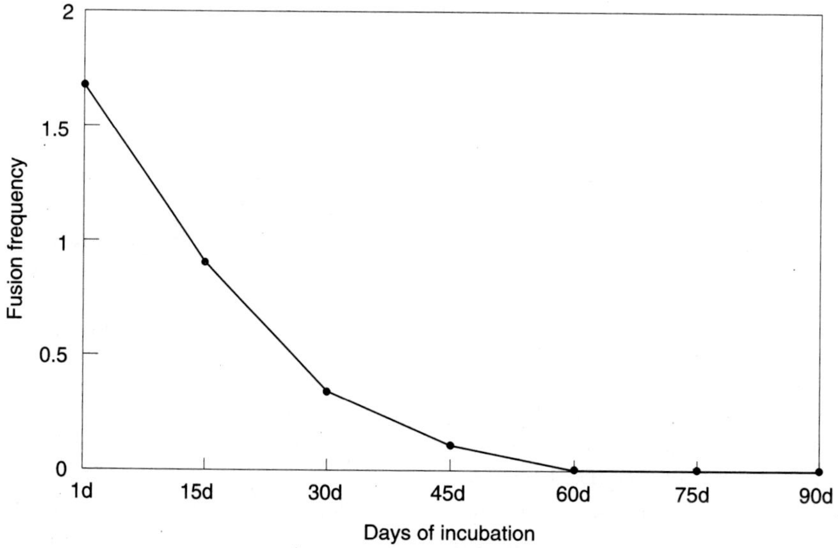

Fig. 25 Effect of storage of protoplasts at 4°C on fusion frequency.
Source: Mrinalini, 1997. Ph.D. Thesis, University of Madras, India, 25-27.

Although membrane fusion is a very important event in living organisms, the underlying molecular mechanism of fusion is not yet completely understood. The pre-requisite for fusion is that the membrane to be fused will be in such close contract that intramolecular interactions can take place between structures that are normally at an intermolecular distance. The barriers separating biomembrane include first of all a repulsive hydration force originating from bound water at the head group of the lipid molecules. When the distance between two opposing membranes is reduced to less than 2-3 nm, a powerful hydration repulsion occurs between hydrophilic surfaces (Rand, 1981). Other barriers of repulsion include plasma membrane (glyco) proteins, macromolecules and other electrostatic repulsion forces (Blumenthal, 1987). To obtain cell aggregation and subsequent fusion the barriers have to be removed or overcome. An external pressure of $10^4 - 10^5$ pa is essential to remove the water layer and establish a close contact between the membranes. The induction of cell aggregation and subsequent fusion vary with different classes of fusogens.

Paramyxoviruses, able to induce cell fusion as mentioned earlier, bind to the cell surface and then fuse with the cytoplasmic membrane under particular conditions. Agglutination is realized by the spike glycoproteins dispersed on the outer site of the enveloped viruses causing cross-bridging of the adjacent cells by means of the virus particles. In Sendai viruses, two major glycoproteins, HN (hemagglutinin and neuramidase activity), and a fusion protein are present on the surface of the envelope. HN is responsible for attaching the virus to cell-surface sialic acid residues covalently linked to glycoplipids and glycoproteins which act as primary cell surface receptor sites (Uchida, 1988). The fusion protein is involved in the fusion reaction itself. A conformational change at the N-terminal sequence of the non-functional precursor fusion protein lies at its origin. As a consequence of this change a stretch of highly conserved and extremely hydrophobic amino acid residues are produced. It is suggested that the hydrophobic stretch of the fusion protein penetrates into the target membrane, provided hydrophobic interaction between the viral membrane and could induce membrane fusion. The mechanism of fusion is still unknown, but several morphological and biophysical studies suggest that as a consequence of the hydrophobic interaction water should be expelled causing a local dehydration at the interbilayer contact sites. This may induce a local transient disordering of the equilibrium bilayer configuration in the two approaching membranes (Burger, 1991) allowing the interaction and intermixing of disturbed lipid molecules of these closely apposed membranes. Different models of intermediate fusion stages have been proposed including the existence of inverted hexagonals configurations (Verkleij 1984; Siegel 1986).

In chemically induced fusion other factors have to be involved. Ca^{2+} are known to dehydrate the space between phospholipid bilayers, they cross-link membranes, and they have also the capacity to neutralize the negatively charged head groups of acidic phospholipids resulting in altered membrane fluidity and phase separation of the bilayer, factors believed to be important in membrane fusion. For PEG, neither aggregation nor fusion do arise from a direct interaction between PEG and the membranes, because PEG is excluded from the area of close contact of the apposed membranes as observed during electron-microscopic investigations (Knutton, 1979). It is believed that the strong dehydrating effect of PEG is the driving force for membrane apposition and is at the origin of altered physico-chemical properties at the vicinity of the membrane surface (Pratsch *et al.*, 1989). Dehydration alone is not sufficient to explain the fusogenic properties of PEG, since the dehydrating agent Dextran does not act as a fusogen, indicating that other factors have to be involved as well. PEG decreases the surface potential of the membranes by several hundreds of millivolts (Maggio *et al.*, 1976), and it may induce alterations in the orientation and hydration of the pho pholipid head groups causing charge neutralization, segregation of the lipids and bilayer defects (Boni *et al.*, 1984) probably inducing a type H_{II} non-bilayer structures between the two interacting apposed membranes. It is assumed that the combination of these effects makes PEG an efficient fusogen.

Also the fusion promoting activity of lysolecithin, monoglycerides and fatty acids is believed to be correlated to the possibility that these compounds promote the formation of H_{II} non-bilayer lipid structures (Hope and Cullis, 1981).

In electro-induced cell fusions the energy of the breakdown pulse provides another means of perturbation of the membrane phospholipids. An alteration in the organization of the polar heads may explain the membrane permeabilization. For electrofusion it has been suggested that there should be no stable intermediate structure. The real fusion intermediate should be a local disorder of lipid molecules that directly leads to membrane rupture after extensive thinning under the influence of the electric forces (Lucy and Akong, 1986).

Fusion is completed by the formation of a small aqueous pore connecting two originally separated aqueous compartments. The extremely unstable configuration of the inverted micelle collapses probably by the system's tendency to reduce the curvature energy of the monolayer (Leikin *et al.*, 1987). The pore is initially about 10-15 nm in diameter, but rapidly widens as swelling under the influence of the osmotic pressure proceeds and finally the contents of the fused cells can be intermingled.

CHAPTER 4

Applications of Protoplast Fusion in Filamentous Fungi

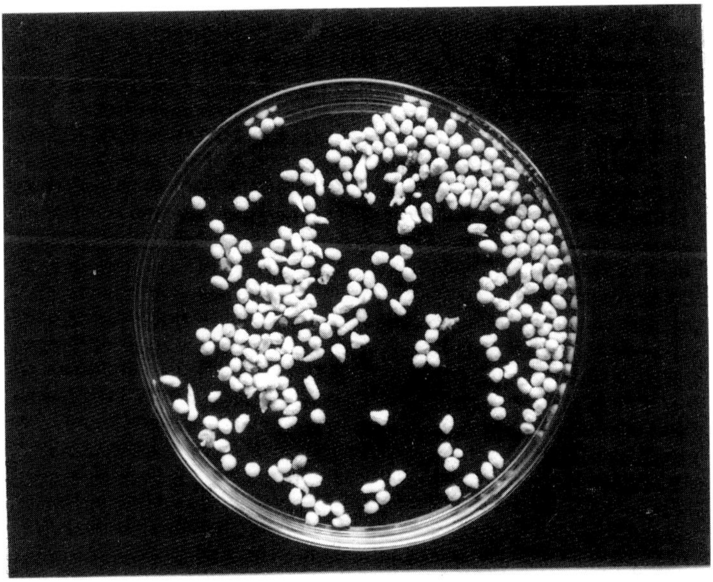

Alginate entrapped super strain of *Trichoderma* spp. fusant

Protoplast fusion is an excellent alternative approach to the construction of fungal strains with novel gene combinations. Although segregants derived, from hybrids as a result of protoplast fusion may represent a new mix of gene pools, protoplast recombination can be used to blend several desirable properties together in a single biotype. This novel technology has not been used intensively. By employing this technique, barriers preventing genetically unrelated fungal strains, even those at present assigned to different genera (e.g. *Gliocladium and Trichoderma*), can be overcome. If mating specificity of potential biocontrol fungi could be completely overcome by partial or complete removal of cell walls, then protoplasts of any two unrelated strains, even those from two fungal genera, should be able to fuse and the fusion products should be stable again. This might indicate that the cell wall constituents may be the only important barrier preventing unrelated fungal taxa from combining their genomes. It would, therefore, be expected that owing to lack of specificity in the process of protoplast fusion (Peberdy, 1983), formation of fusion products between protoplasts of *Trichoderma* and *Gliocladium*, or *Penicillium vermiculatum* (*Talaromyces flavus*) and other *Penicillium* spp. will be possible. Ferenczy *et al.* (1975 b) described a simple technique that results in high frequency interspecific protoplast fusion and postulated that among the tested organisms the results of *Aspergillus-Penicillium* protoplast fusion, would be especially interesting from both theoretical and practical points of view. In the same context *Glioderma* strains combining the enhanced biocontrol ability of *Gliocladium* and tolerance to benomyl from *T. viride* or *T. harzianum* mutants could theoretically be produced as a result of *Gliocladium* and *Trichoderma* protoplast fusion. Protoplasts of *Saccharomyces lipolytic* and *Pichia guilliermondie* (*Spata* and *Weber*, 1980), and those of *S. cerevisiae* and *S. pombe* (Svoboda, 1980) have been used succesfully though the hybrids were unstable.

4.1 PROTOPLAST FUSION RELATIONSHIP TO PARASEXUAL CYCLE

The Deuteromycetes or Fungi Imperfect do not possess a sexual stage. How then do these fungi generate variability to adapt to an ever changing environment if all that is available to them is mitotic divison? The answer was presented by Pontecorvo and Roper (1952) who discovered parasexuality in *A. nidulans*. Since then, parasexual phenomenon has been observed in several imperfect fungal species as well as in Basidiomycetes and Ascomycetes other than *Aspergillus* (Alexopoulos and Mims, 1993; Hastie, 1981). This process is defined as genetic recombination in which there is no definite coordination between segregation, recombination and chromosome reduction at meiosis. Alternatively, it is defined as a cycle in which plasmogamy, karyogamy and haploidization take place but not at specified time, places, or points in the life cycle of an organism. Functionally, the parasexual cycle can be divided into a series of events (Pontecorvo, 1956; Hastie, 1981).

a. Fusion of protoplasts or hyphal anastomosis to form the heterokaryotic mycelium (plasmogamy)
b. Infrequent nuclear fusion (Karyogamy) to form both homozygous diploid and heterozygous diploid nuclei

c. Multiplication if diploid nuclei side by side with the unfused haploid nuclei
d. Occasional mitotic crossing over during multiplication of the diploid nuclei followed by mitosis with occasional nondisjunction to form a heterokaryon with parental and recombinant genotypes.
e. Nuclear segregation leading to homokaryosis.

Protoplast fusion generates mostly the same type of conditions as in the parasexual cycle. A number of events occur once two protoplasts fuse. While the fusion product reverts to normal form, the nuclei remain separate and divide mitotically forming a heterokaryotic colony (Ferenczy *et al.,* 1975 b; 1984. Anne *et al.,* 1976; Anne, 1977; Kevei and Peberdy 1984;) The two genetically different nuclei in the heterokaryon may fuse, forming a stable or unstable heterozygous diploid colony or sector (Svoboda, 1977; Kevei and Peberdy, 1984; Raymond *et al.,* 1986). These diploid colonies can then spontaneously lose chromosomes to achieve an aneuploid or haploid state or the colonies can be induced to undergo haploidization by treatment with para-fluorophenylalinine, of benomyl, UV or X-rays. (Sipiczki and Ferenczy, 1977; Peberdy *et al.,* 1977; Sarachek *et al.,* 1981).

4.2 MATING TYPE AND INCOMPATIBILITY GROUP BARRIERS

Protoplast fusion has been applied to organisms containing mating type and incompatibility group barriers. As a result, genetic recombination experiments can be done between sexual or heterokaryon incompatible strains. In *Saccharomyces cerevisiae* mating normally occurs between heterothallic strains of the opposite mating type. Van Solingen and Van der Plaat (1977) found that fusion of two strains of *S. cerevisiae* of mating type a resulted in fusion products that were diploids of mating type a. Savoboda (1978) fused two strains of mating type a and found the fusion products to be of mating type a. Other investigators obtained similar results with *S. cerevisiae* (Ferenczy and Maraz, 1977; Svoboda, 1977; Gunge and Tamaru, 1978; and with other organisms such as *Schizosaccharomyces pombe.* Sipiczki and Ferenczy (1977) fused two strains of *S. pombe* of mating type 'h' and found the resulting fusion products were diploid and would mate only with h+ strains. Similar results have been reported for *Rhodosporidium toruloides* (Sipiczki and Ferenczy, 1977) and *Saccharomycopsis lipolytica* (Stahl, 1978).

The protoplast fusion technique has also been used to successfully fuse *Aspergillus nidulans* protoplasts of different heterokaryon compatibility (h-c) groups (Dales and Croft, 1977; Croft, 1985), a condition that normally would not occur in nature. This fusion event leads to recovery of both heterokaryons and diploids. A linkage map of genes responsible for h-c groups in *A. nidulans* has been constructed by fusion of h-c incompatible strains and recovery of recombinants by treatment of diploids with benomyl (Croft, 1985). Typass (1983), while studying *Verticillium albo-atrum* and *V. dahliae*, found that a number of incompatibility factors exist in *Verticillium* that prevent formation of heterokaryons via normal parasexual processes. However, application of the protoplast fusion technique resulted in the isolation of both heterokaryons and diploids. The same type of analysis has recently been carried out in *V. lacanii* using protoplast fusion (Jackson and Heale, 1987). There are contrasting examples where protoplast fusion techniques will not allow formation of viable fungi across heterokaryon incompatibility barriers (Kevei and Peberdy, 1977; Adams *et al.,* 1987).

4.3 SPECIES AND GENUS BARRIERS

Protoplast fusion can also be used to overcome species and genus barriers producing interspecific and intergeneric hybrids. The fusion frequency, an indicator of the success of a fusion, often falls 2-6 orders of magnitude below that of an intraspecific fusion and is probable indication of the amount of genetic relatedness between the parent species involved in the fusion.

Early studies suggested that hybrid formation with subsequent generation of non-parental types could be achieved only if closely related parental species were used. Fusion products between *A. nidulans* and *A. rugulosus*, closely related species of the same species group, first grow as heterokaryotic colonies on minimal medium. Sectors of more vigorous growth arise from the heterokaryons and on the basis of segregation patterns and DNA estimation, were deemed to be interspecific hybrids whose origin resulted in fusion of the two parental nuclei. These fusants were shown to be diploid and/or aneuploid in nature (Kevei and Peberdy, 1977; 1979; 1984; 1985; Bradshaw *et al.,* 1983). Crosses between *P. chrysogenum* and *P. cyaneo-fulvum,* two closely related species, yielded colony types that behaved similarly (Peberdy *et al.,* 1976). As a matter of terminology, interspecific or intergeneric diploids are referred to as allodiploids while aneuploids are referred to as interploids, partial alloploids and partial heteroploids.

4.4 MITOCHONDRIAL TRANSFER USING PROTOPLASTS

Protoplast fusion technique is used to study nucleus-mitochondria interactions and mitochondrial genetics and biogenesis. Intraspecific nonselective mitochondrial transfer was accomplished by Ferenczy and Maraz (1977) in *S. cerevisiae.* One strain had the genotype of a ural trp 1 (ER) and the other strain ade2 (rho o ER). The fusion products were characterized as respiratory competent and uninucleate; they had the expected diploid DNA level and were erythromycin-resistant. Other studies have also shown that transmission of mitochondria by protoplast fusion is possible and that recombinational processes involved in mitochondrial genetics are independent of thc mating type barriers. Mitochondrial markers have been used extensively in yeast strain improvement programmes in industry. These yeasts are typically diploid or of higher ploidy (Stewart, 1981). Introduction of markers in the nuclear genome is therefore difficult (Stewart, 1981); and often causes deleterious effects in vigor and stability of production strains as well as production losses (Fantini, 1962; MacDonald *et al.,* 1963; Stewart, 1981; Ferenczy, 1984; Mellon, 1985). On the other hand, both mitochondrial petite and resistant mutations can readily be introduced and applied in hybrid formation (Spencer and Spencer, 1983).

Non-selective mitochondrial transfer in the petite-negative yeast *Kluyveromyces lactis* has proven to be the only genetic method available to study respiratory-deficient mutants of this yeast (Allmark *et al.,* 1978). The technique has also proven useful in studying transmission and recombination frequencies as well as segregation patterns of mitochondrial markers in *Schizosaccharomyces pombe*. The first attempt at controlled intraspecific selective mitochondrial transfer or the transfer of isolated mitochondria (Gunge and Sakaguchi, 1979) utilized a mutant strain of *S. cerevisiae* carrying oligomycin resistant mitochondria. These mitochondria were isolated and fused with another strain of *S. cerevisiae* which has rho O and carried different nuclear markers. The frequency of transfusion was less than 1×10^8/ regenerated protoplast. Several investigators have since obtained higher frequencies (Yoshida, 1979; Yoshida and Takeuchi, 1980).

Interspecific non-selective mitochondiral transfer was carried out with strains of *A. nidulans, A. nidulans* var. *echinulatus* and *A. quadirlineatus* (Croft, 1985). The molecular weights and restriction patterns of the mitochondrial DNA were obtained with high frequency when the small mitochondrial genome was selected for transfer to the species containing the arger genome. After fusion of *A. nidulans* and *A. nidulans* var. *echinulatus* protoplasts, it was found that the mitochondrial DNA differed only in a small number of species-specific restriction sites and in six insertions/deletions (Earl *et al.,* 1981). Petite mutants of *S. cerevisiae* have been used as fusion partners with wild type strains of *Candida pseudotropicalis*, *Hansenula anomala, H. Capsulata, S. montanus, S. rosei,* and *Schizosaccharomyces pombe* resulting in respiratory competent fusion products (Spencer & Spencer, 1980). Interspecific selective mitochondrial transfer using respiratory deficient *S. cerevisiae* protoplasts as recipients for mitochondria from *Hansenula wingei* and *Schizosaccharomyces prombe* have been reported.

Restoration of the rho+ character was reported following uptake of isolated forieng mitochondria (Yoshida, 1979; Yoshida and Takeuchi, 1980) or with miniprotoplasts containing mitochondria.

4.5 A GENETIC EXCHANGE IN DEUTEROMYCETES AND ASCOMYCETES

There are many fungi of pathogenic, economic and medical importance, which are members of the class Deuteromycetes, e.g. *Candida* species. This class of fungi do not possess a sexual stage (Alexopoulos and Mims, 1993) and therefore, conventional crosses and selection of desirable progeny are not an option with these fungi. Protoplast fusion has been demonstrated in *C. albicans* (Vallin and Ferenczy, 1978; Sarachek *et al.,* 1981; and *C. utilis* (Delgado and Herrera, 1981) and has also been successful in filamentous Deuteromycetes. For example, *Beauveria bassiana,* an entomopathogen with biological control applications, has been fused (Kawamoto and Aizawa, *1986; Other* members of the genus *Beauveria* have also been fused. Both intraspecific and interspecific fusions have been reported in *B. amorpha, B. bassiana* and *B. Brongniarti* (Shimizu, 1987).

The ascomycetes such as *Aspergillus* and *Penicillium* species, may have a sexual cycle but it is difficult and sometimes impossible to demonstrate (Alexopoulos and Mims, 1993). Protoplast fusion of either Deuteromycetes or Ascomycetes with different genetic characteristics offers the possibility of isolating fusion products with the desired character.

4.6 PROTOPLAST FUSION IN STRAIN IMPROVEMENT

Protoplast fusion is of current interest because of its applications in pure and applied genetics. Protoplast fusion technology is applied for developing intraspecific, interspecific and intrageneric suprahybrids with higher potentiality than their parent strains. Through protoplast fusion technique, improved strains with enhanced antagonistic potential, antibiotics, enzymes, useful mycoproducts, high yielding mushrooms, etc. could effectively be achieved. Many of the unsolved problems of pathogenesis can be unveiled through protoplast fusion. The utmost importance and chemistry of cytoplasmic inheritance of phytopathogenic fungi can clearly be analysed through protoplast fusion. The mode of acquired resistance, adaptability, survival of fitness and build up of fungicide resistance under natural environment could clearly be understood through protoplast fusion. Protoplast fusion helps in overcoming vegetative incompatibility and helps in hybridization. Though nuclear hybridization is spontaneous, the mitochondrial recombination is a common feature in improved strain with enhanced potentials.

Biocontrol of the plant pathogen is increasingly becoming an important and valuable method of plant disease management. *Trichondrma* spp. have a potential to become industrially important fungi, both as sources of useful enzymes (cellulase, pectinase, chitinase) and as antagonist of plant pathogens. Enhanced antagonistic potential of biocontrol agents is important for the success in control of plant diseases and their ultimate acceptance in integrated disease management. Genetic manipulation of biocontrol agents especially *Trichoderma* spp. is essential for strain improvement. Manipulation by sexual cycle and anastomosis is not so successful. Protoplast fusion is an effective alternate technique for studying the genetics and biochemistry of fungi leading to new approaches in genetic manipulation (parasexual hybridization).

a. Enhanced Efficacy of Biocontrol Agents

1. Protoplast fusion and strain improvement in Metarhizium anisopliae (Silveria and Azevedo, 1987)

The Deuteromycete *Metarhizium anisopliae* (Metsch) Sorokin is an entamopathogenic fungus which is being used commercially as biological control of insect pests mainly those which attack sugarcane and gramminaceous crops in Brazil. Using two strains of *M. anisopliae var minor* designated E6 and R1 from different origins, interstrain crosses were readily performed by orthodox methods from mutants of strain E6 through the parasexual cycle. However, in most cases crosses between mutants of strains were not achieved. Protoplast fusion was carried out using PEG as fusogen. Sectors which emerge from fusion products were analysed and recombinants were obtained even from crosses between mutant strains which could not be crossed by hyphal fusion. In this way protoplast fusion proved to be a valuable tool for further studies of genetics and breeding of *M. anisopliae*.

2. *Strain improvement of Trichoderma* spp.

Protoplast fusion between two species of *T. harzianum* and *T. longibrachiatum* to bring together the characters of the parent, i.e. the former for antagonism and latter for tolerance to $CuSO_4$ and carbendazim has been successfully exploited (Lalithakumari *et al.,* 1996). Protoplast fusion increased the efficacy of the biocontrol strain through enhanced antagonistic potential intergrated with acquired tolerance to fungicides against a broad spectrum of phytopathogens. The fusants (Color Plate 4) exhibited characters of both the parents resembling *T. harzianum* in morphological character like growth pattern and pigmentation and *T. longibrachiatum* in tolerance to carbendazim and $CuSO_4$ (Colour Plate 5) and the spore size was intermediate to both parent strains (Table 30).

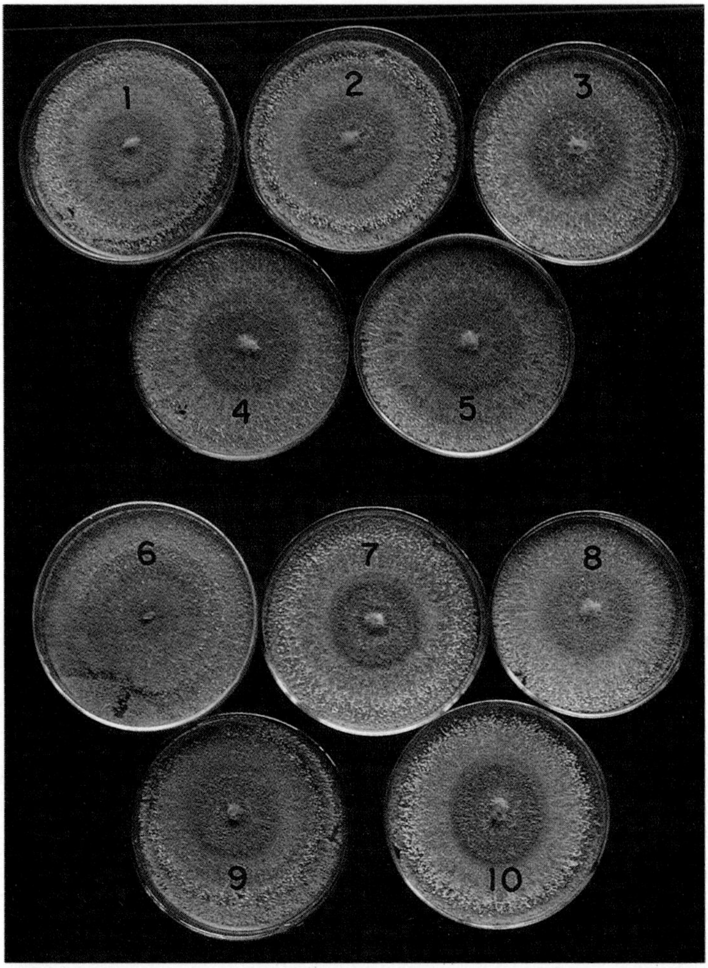

Colour Plate 4 Interspecific fusants between *T. harzianum* and *T. longibrachiatum* (1-10)

Table 30 The measurement of hyphae, phialides and spores of parent and fusant strains of *Trichoderma*

	Strains of *Trichoderma*		
	T. harzianum 1	*T. longibrachiatum*	Fusants 1&2
Hyphae μM	1.5-2	2.5-10	2.5-8
Phialide μM	3-3.5 × 5-7.5	2.5-3.5 × 6-15	2.5-3.5 × 6-15
Spore μm	2.5-3 × 2.8-3	2.5-3.5 × 6	2.5-3.5 × 6
Pigmentation	Dull Green	Olive Green	Whitish to Dull Green

Source: Mrinalini, 1997. Ph.D. Thesis, University of Madras, India 38-40.

Spore size or conidiophore diameter has also been used in most of the interspecific fusion (Peberdy *et al.*, 1977; Anne *et al.*, 1982). The self fusion products of each parent were kept as control. Both the parental fusants and hybrid fusants were stable for 4 generations. The total DNA content in the parent and fusant strains when estimated showed a slight increase of 3-8% in both the fusants per conidium (Table 31). The mitochondrial restriction enzyme fragments showed typically intermixed identical bands between them and between either parents (Plate 18). The frequent occurrence of non-parental restriction fragments in the mitochondrial DNA of the progeny of a cross between two strain indicates the high frequency with which mitochondrial recombination takes place. Protoplasts fusion facilitates the transfer of mitochondrial genomes between taxonomically related but quite distinct species.

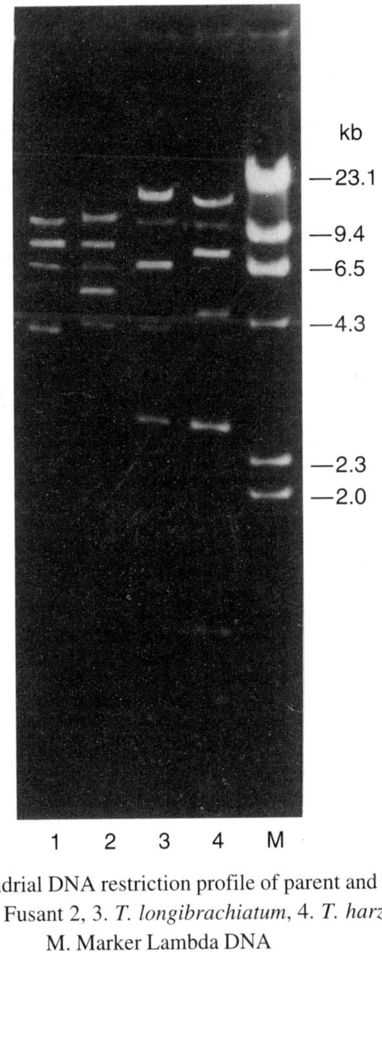

Plate 18 Mitochondrial DNA restriction profile of parent and fusant strains 1. Fusant 1, 2. Fusant 2, 3. *T. longibrachiatum*, 4. *T. harzianum*, M. Marker Lambda DNA

a b c

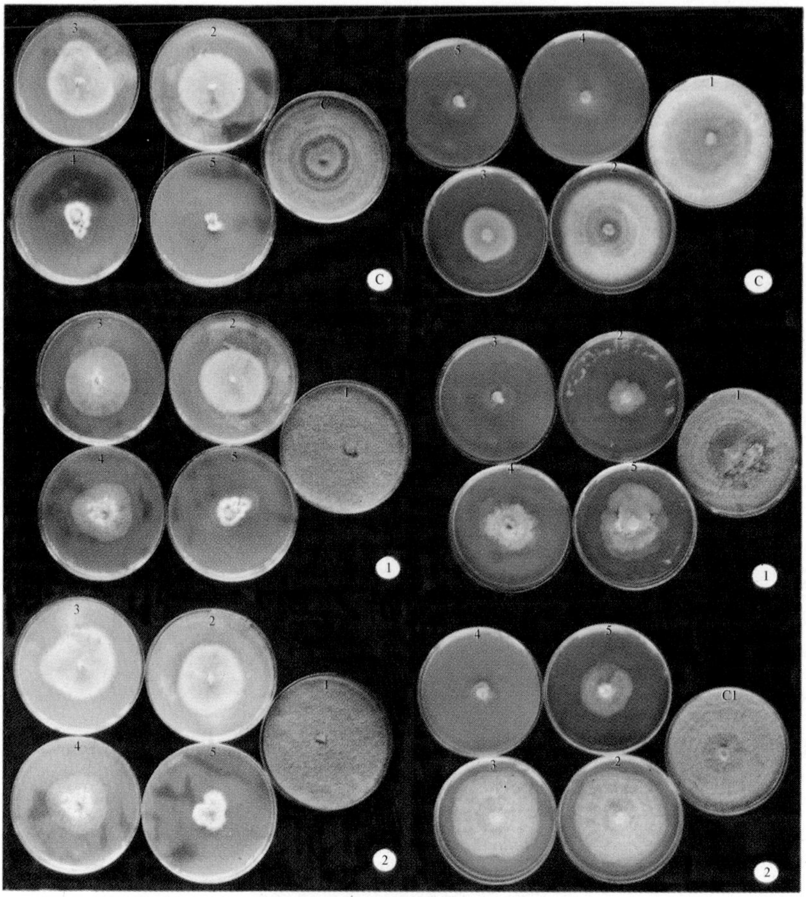

a b c

CuSO$_4$ (A)

Carbendagim (B)

Colour Plate 5. A. Growth of parent and fusant strains of *Trichoderma* on copper sulphate B. Growth of parent and fusant strains of *Trichoderma* on carbendazim. a. *T. longibrachiatum*; b. Fusant 1; c. Fusant 2; 1. Control; 2. 50 mM; 3. 100 mM; 4. 150 mM; 5. 200 mM.
Source: *Mrinalini, 1997.* Ph.D. Thesis, University of Madras, India, 92-94.

Table 31 The total DNA content and GC % of DNA of parent and fusant strains of *Trichoderma*

Strains of *Trichoderma*	DNA content (μg/mg of conidia)	GC%
T. harzianum 1	5.53	59.9
T. longibrachiatum	6.84	63.1
Fusant 1	6.95	66.2
Fusant 2	7.11	71.3

Source: Mrinalini, 1997 Ph.D. Thesis, University of Madras, India 95-96.

The fusants with enhanced antagonistic efficacy integrated with tolerance to fungicide carbendazim and $CuSO_4$ proved as best biocontrol agent in pot culture experiments. The fusants also established their saprophytic competency when tested under pot culture conditions with unsterilized soil (Color Plate 6). Paddy plants treated with the fusants showed an increased growth, tillering and increased height when compared to other treatments. The fusants entrapped in alginate beads (immobilised) were mixed with paddy seeds for better results.

Colour Plate 6 Green house experiment to prove the efficiency of parent and fusant strains of *Trichoderma* in the control of sheath blight disease of paddy. 1. Infected with *R. solani;* 2. Treated with fusant 1; 3. Treated with fusant 2; 4. Treated with *T. longibrachiatum;* 5. Treated with *T. harzianum.*

Source: *Mrinalini, 1997.* Ph.D. Thesis, University of Madras, India, 100-102.

4.7 USE OF AUXOTROPHIC MUTANTS IN PROTOPLAST FUSION

Recognition or selection of fusion products is based upon the prototrophic or partially complemented nature of the resulting colonies and in addition, the fusion frequency may be calculated by comparing the number of colonies developing in nutritionally incomplete medium to that growing in a complete one. Of course, the frequency of protoplast fusion and the frequency of nutritional complementation can never be the same since fusion also occurs between identical (non-complementary) partners.

Whatever methods are used for protoplast formation and fusion one of the final tasks is to characterize the fusion products and to determine the nature of nutritional complementation.

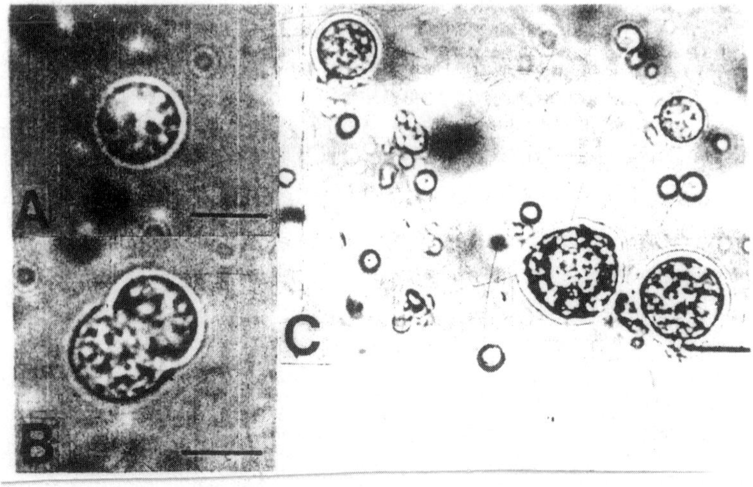

Plate 18a Mycelial protoplast of *T. reesei* QM 9414 and *T. reesei* QM 9136. A free protoplast (after 5 h).
Reproduced from Ogawa et al., 1989. Ferment. B. Oeng. 67: 207-209.

In the past few years a considerable mass of data has been accumulated on variations in the genetic background of nutritional complementation as a consequence of protoplast fusion.

a. Heterokaryon with Frequent Stable Diploid Formation

A characteristic species displaying this phenomenon is *Candida tropicalis*. From auxotrophic mutants heterokaryons, heterozygous prototrophs, somatic segregants and recombinants could be obtained. Interestingly, uninucleate prototrophs originating from diauxotrophs spontaneously gave rise to monoauxotrophs. This might indicate the existence of aneuploids. This idea is also supported by the low DNA content of certain prototrophs. If red adenine-requiring and white cysteine multinucleate heterokaryotic prototrophs were coloured a wide variety of shades of pink, these readily gave rise to auxotrophic cells of the parental types even on minimal medium. The diploid (and/or aneuploid) prototrophs were white, rapidly growing, and genetically extremely stable.

b. Stable diploid formation

Using haploid auxotrophs of identical mating type, stable diploids could be constructed in a series of yeast species. Heterokaryon formation was so transient that these were not detected the diploids were stable and usually spores were not produced. The fusion cells were uninucleate, enlarged, and their DNA content was about twice that in the parental haploid cells. By induced haploidization or by crosses with cells of the opposite mating types, both haploids with the parental markers and recombinants could be recovered. If protoplasts of opposite mating types were fused, the fusion products showed the characteristics of normal crosses.

In addition to diploids, cells of lower and higher ploidy and different types of aneuploids were also occasionally observed as a result of the fusion of haploid protoplasts. Even multiple fusion leading to triploids and tetraploids may be encountered. Triploid fusion products have been formed by protoplast fusion of two different strains as well as three different strains.

c. Heterokaryon and Heteroploid Formation

If protoplasts to be fused are derived from more distantly related species, the genetical background of nutritional complementation will be more complex. Heterokaryon formation is the first event, and the heterokaryotic state can be stable. Synkaryosis is also possible, leading to complicated genetic situations not existing in nature. One can fully agree with Dales and Croft (1977) that instead of the intrinsically homospecific terms 'diploid' or 'aneuploid', 'heteroploid' and 'partial heteroploid' should preferably be used in these heterospecific cases.

Such a situation has been observed in the case of the somatic hybrids of *Penicillium roquefortii* and *P. chrysogenum.* After protoplast fusion of auxotrophic mutants slowly growing prototrophic colonies developed which could be classified into three types. Type 1 colonies were morphologically normal, grew faster on nutritionally rich media than on minimal medium and produced selectively auxotrophic *P. roquefortii* conidia on rich media. Type 2 colonies were morphologically aberrant in consisting a loosely meshed network of broadly spreading hypahe. They also produced selectively auxotrophic *P. roquefortii* conidia on rich media. In contrast to these, type 3 colonies sporulated on minimal medium, released large prototrophic conidia, and were similar to *P. chrysogenum* in morphology. All types produced penicillins of the same chemical composition as those of *P. chrysogenum.* Type 1 and 2 colonies were heterokaryotic, whereas type 3 might have been either full or partial heteroploid.

Interspecified hybrids of auxotrophic mutants of *P. citrinum* and *P. cyaneo-fulvum* produced slowly growing colonies on minimal medium and conidia of both species were released. More vigorous developing sectors appeared in these colonies with white, large and stable prototrophic conidia which were assumed to contain the complete set of chromosomes of both the complementing species. Markers of both species in almost equal numbers were found on induced segregation. Among the segregants partial heteroploids could also be detected.

No details are known about the genetic background of interspecific complementation after protoplast fusion of auxotrophic strains of *Candida tropicalis* and *Saccharomycopsis fibuligera.* Cells of the prototrophic fusion products proved uninucleate and exhibited assimilation spectra resembling those of *Candida* or *Saccharomycopsis* or both. The hybrids were perhaps, partial heteroploids.

Recognition of selection of fusion products is based upon the prototrophic or partially complemented nature of the resulting colonies and, in addition, the fusion frequency may be calculated by comparing the number of colonies developing in nutritionally incomplete medium to that growing in a complete one. Of curse, the frequency of protoplast fusion and the frequency of nutritional complementation can never be the same since fusion also occurs between identical (non-complementary) partners.

Whatever methods are used for protoplast formation and fusion one of the final tasks is to characterize the fusion products and to determine the nature of nutritional complementation.

In the past few years a considerable mass of data has accumulated on variations in the genetic background of nutritional complementation as a consequence of protoplast fusion.

d. Partial Heteroploid Formation (Bidirectional)

If protoplasts of auxotrophic cells of the distantly related species of *Aspergillus nidulans* and *A. fumigatus* were fused, slowly growing abnormal colonies were produced with a characteristically low frequency; the interspecific fusion frequency was at least 5 orders of magnitude less than that of the intraspecific one in either species. The heterokaryotic or full heteroploid states were so transitory that heterokaryons or full heteroploids (interspecific diploids) could not be isolated. Presumably, both the heterokaryotic and the diploid states are lethal. the heteroploid state may exist only a genome of one of the parents and only one or a few chromosomes from the other. On nutritionally rich medium the complemented cells rapidly segregated one of the partners. The segregation was bidirectional, since either the parental. A. *nidulans* or *A. fumigatus,* but never both, could be recovered.

e. Partial Heteroploid Formation (Unidirectional)

Protoplasts of stable monoauxotrophic strains of *Kluyveromyces lactis* and *K. fragilis* were successfully fused. The fusion colonies were able to maintain their prototrophy even in complete medium. In general, the fusion cells were larger and contained more DNA than those of either of the parent species. On the other hand, in most cases the DNA content was lower than the combined DNA content of the two parent implying that subsequently to fusion loss of chromosomes had occurred. At the same time, a selective retention of *K. fragilis* and loss of *K. lactis* mitochondrial DNAs was observed. Cells of the prototrophic fusion products proved uninucleate and exhibited assimilation spectra resembling those of *Candida* or *Saccharomycopsis* or both. The hybrids were, perhaps, partial heteroploids.

The industrial application of fungi are considerable and very diverse. Different compounds are produced by these organisms, from low molecular weight primary metabolites such as ethanol or organic acids to secondary metabolites such as alkaloids and antibiotics.

In experiments aiming at fusion of fungal protoplast from auxotrophic parental mutants in most cases are based on nutritional complementation indicative of protoplast fusion. Auxotrophic mutant strain leucine$^-$ from *T. reesei* and methionine from *T. harzianum* were used for fusion studies. The genus *Trichoderma* is well known for their lytic enzyme production. *T. harzianum* produces chitinase xylanse and trace amounts of β-1, 3 glucanase and cellulase while *T. reesei* produced cellulase, xylanase and trace amount of chitinase. The feasibility of protoplast fusion between *T. harzianum* (meth$^-$) and *T. reesei* (leu$^-$) for the production of cellulase, chitinase and β-1, 3 glucanase of this interspecific fusant has been worked out (Chellappa, 1997).The auxotrophic mutants produced were characterized and compared for their growth and ability to produce chitinase, cellulase, β-1-3-glucanase and xylanase with the parent strains. There was change in growth rate, morphology, pigmentation, sporulation and their production of lytic enzymes. Further these auxotrophic mutants were used for releasing protoplasts using the optimized age, enzyme mixture, incubation time, pH and osmotic stabilizer reported earlier for the parent strain. Care was taken in such a way that the fusant studies were carried out immediately after releasing protoplasts from the auxotrophic mutants since storage of protoplast was also a crucial factor in success of fusion frequency.

PEG 40% (w/v) concentration was effective in the presence of 0.6 M KCl as osmotic stabilizer with 10 mM $CaCl_2$ and 10 mM Tris-HCl in bringing about a very good and high fusion frequency and the optimum pH for the fusion was 7.5 at normal room temperature (28°C) and also it was observed that mixture of protoplasts of both the strains did not form colonies without centrifugation.

Although the auxotrophic parental mutant shows less enzyme activities, the heterokaryon showed enhanced enzyme (named as CHELKZYME) production and it showed both the parental enzyme, i.e. cellulase and chitinase activities, as confirmed by ELISA technique (Plate 19).

The purified enzyme screened (CHELKZYME CASB) induced the release of protoplasts from all the fungal species examined. The efficacy of Chelkzyme with NOVOZYM 234 and Onozuka when compared showed

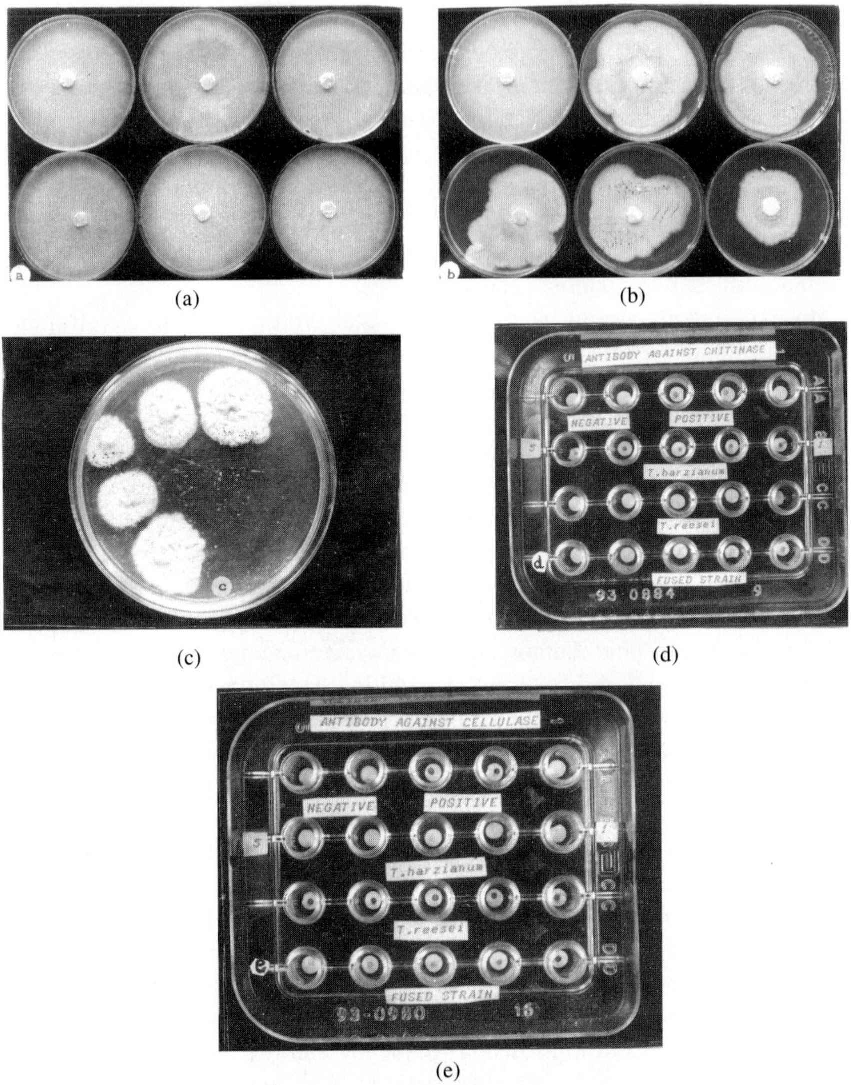

Plate 19 a. Auxotrophic mutant of *T. harzianummeth;* b. Auxotrophic mutant of *T. reesei* (leu); c. Fusants of *T. harzianummeth* and *T. reesei* (leu); d.e. Screening of fusants for chitinase and cellulase production by ELISA technique.

Source: Chellappa, 1997. Ph.D. Thesis, University of Madras, India, 66-69.

higher protoplast release from fungi and plants. Among the fungi tested, the highest production of protoplasts, over 1×10^7/ml was obtained from species of Basidiomycotina; only a few were obtained from species of Zygomycotina. The yield from Ascomycotina and Deuteromycotina varied depending on the species.

4.8 INTRASPECIFIC HYBRIDIZATION OF *TRICHODERMA REESEI* BY PROTOPLAST FUSION (OGAWA *et al.*, 1989)

The fungus *Trichoderma reesei* QM 9414 is well known as the most suitable organism for the production of cellulase system capable of degrading native cellulosic substances such as an absorbant cotton or Avicel. This fungus also actively produced mycolytic enzymes such as β-1, 3-glucanase and chitinase. The technique of protoplast fusion is potentially an important clue for the genetic manipulation of cellulolytic fungi and hence strain improvement and is being investigated in several laboratories. Protoplast obtained from mycelia of a single auxotrophic mutant of *T. reesei* QM 9414 were fused with those of *T. reesei* QM 9136 (Plate 18a).

The fused protoplasts successfully formed the diploids by treatment with 0.1% d-camphor. Cellulase activities such as filter paper degrading and CMC and Avecil saccharifying activities and the xylanase activities of the diploid showed intermediate values between *T. reesei* QM 9414 and *T. reesei* QM 9136. However the β-glucosidase, β-1, 3-glucanase and chitinase activities of the diploid increased to levels equal to or above those of parent strains, thus confirming the possibility of enhanced enzyme production using protoplast fusion technique.

4.9 PROTOPLAST FUSION OF FUNGICIDE RESISTANT MUTANTS

Protoplast fusion is the most useful technique to understand acquired fungicide resistance. Fungicide resistance ability can be transferred between strains by protoplast fusion. Two fungicide resistant mutants of *V. inaequalis* were fused to prove the probability of development of multiple resistant strain under *in vitro* condition (Color Plate 7).

Of the two mutants, one was resistant to penconazole designated as PRM and another was resistant to carbendazim, (a benzimidazole compound). The fusion frequency with PEG was 60%. The fusants were screened on a range of PDYEA amended with penconazole, with carbendazim and with both penconazole and carbendazim (PDYEA + penconazole, PDYEA + carbendazim, PDYEA + penconazole + carbendazim). The fusants grew on the medium amended with both the fungicides. Penconazole + carbendazim resistant fusants exhibited either parental and both parental (biparental) character. Fusants with biparental characters are considered as recombinants. The fusants were slow growing and on further subculturing, they segregated into fast growing and slow growing colonies. The slow growing colonies

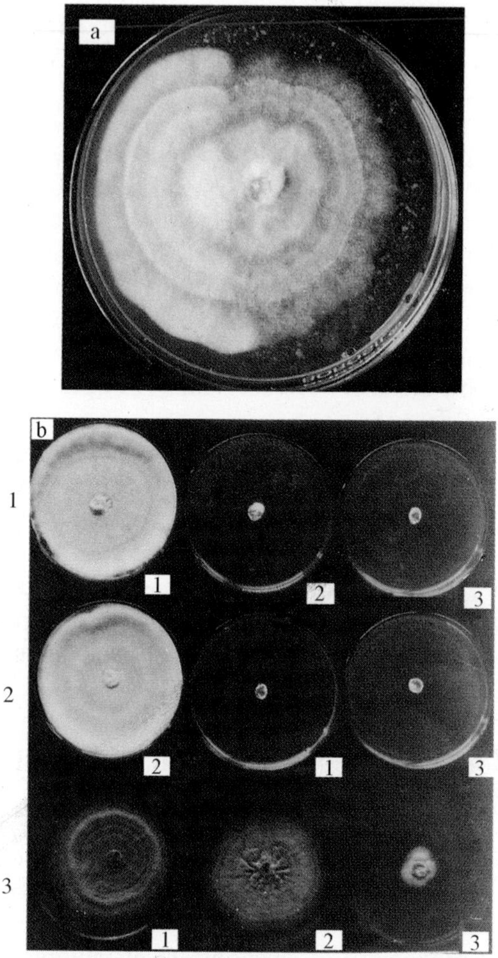

Color Plate 7 Analysis of multiple resistance in the protoplast fusants of *V. inaequalis*

a. Penconazole and Carbendazim resistant mutant

b. 1. Penconazole resistant parent on (i) Penconazole + PDYEA (ii) Carbondazim + PDYEA (iii) Penconazole + Carbendazim + PDYEA

2. Carbendazim resistant parent on (i) Carbendazim + PDYEA (ii) Penconazole + PDYEA (iii) Penconazole + Carbendazim + PDYEA

c. 3. Penconazole and Carbendazim resistant mutant on (i) Penconazole + PDYEA (ii) Carbendazim + PDYEA (iii) Penconazole + Carbendazim + PDYEA

Source: Vijayapalani, 1995. Ph.D. Thesis University of Madras, India. 222-225.

completely lost the pigmentation and the number of spores varied among the fusants. The self fusion products of each parent were kept as control. Both the parental fusants and hybrid fusants were stable for 4 generations. Thus protoplast fusion provided ample information about the transfer of fungicide resistance in plant pathogens and it is an important clue for build up of resistance in the field even after the withdrawal of stress chemical.

4.10 PROTOPLAST FUSION BETWEEN NON-SPORULATING AND SPORULATING STRAINS OF *V. INAEQUALIS*

Protoplast fusion (Lalithakumari, 1994) was carried out between sporulating carbendazim sensitive strain and non-sporulating carbendazim resistant mutant of *V. inaequalis* using PEG. After fusion, the progenies were screened on carbendazim amended medium for sporulating carbendazim resistant fusants. Few colonies could sporulate and showed tolerance to carbendazim (Color Plate 8).

The fusion frequency was determined as 0.1%. The self fusion products of each parent was kept as control. Both the parental fusants and hybrid fusants were stable for four generations. It is another evidence for character transfer through protoplast fusion.

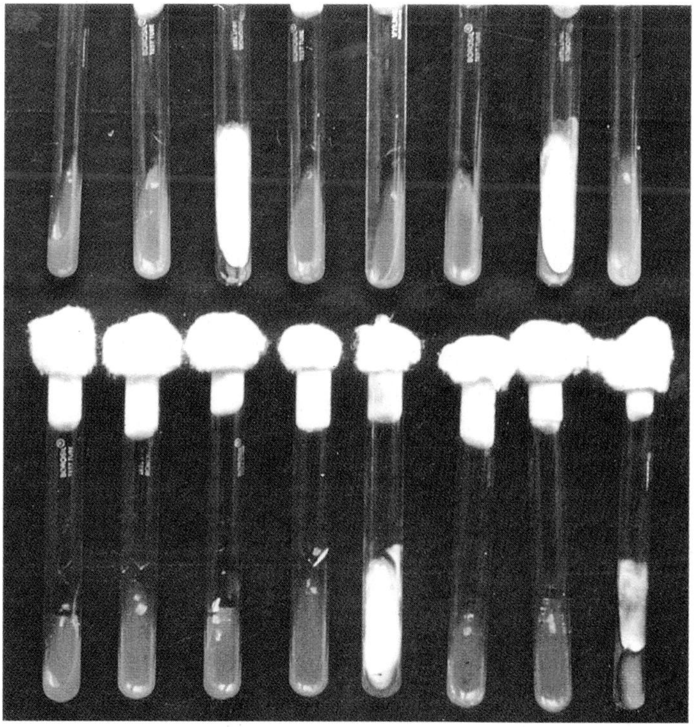

Color Plate 8 Sporulating carbendazim resistant fusant of *V. inaequalis*
Source: Lalithakumari, I. C. A. R. Report. 1993

4.11 PROTOPLAST FUSION IN *PENICILLIUM* (ANNE, 1993)

In an intensive large-scale industrial strain development programe of *P. chrysogenum*, a stable heterozygous diploid strain that produced high concentrations of phenoxymethyl penicillin was isolated and reported as having been used by Eli Lilly for commercial penicillin production. However, stable diploids are rather exceptional in both productivity and strain stability. But, diploids can be used as a source to obtain haploid recombinants with more attractive features than their progenitors. This approach could even be more successful than attempting to isolate high yielding diploids. Isolation of a spontaneous segregant that produced nearly 25% more penicillin than its diploid parent which itself yielded better than its ancestor. In addition, from crosses carried out by protoplast fusion between a slow growing *P. chrysogenum* strain producing hydroxypenicillin V and a strain with faster growth but producing moderate levels of penicillin V and high yields of p-hydroxypenicillin V, recombinants could be recovered with the desired properties of fast growth and high levels of penicillin V, but low p-hydroxypenicillin V production (Lowe and Elander, 1983).

These examples clearly showed the possibility of selecting improved recombinants after segregation of diploids obtained following protoplast fusion. However, in the isolation of recombinants, difficulties were often encountered, not only to obtain stable recombinants, but segregants were mainly of one or another of the haploid parental type. This phenomenon, known as parental genome segregation is probably due to differences in chromosomal morphology between the haploid parental strains as a result of translocation caused by multiple mutagen treatment. The occurrence of chromosomal rearrangements in production strains has in the meantime been physically demonstrated for *Cephalosporium acremonium* by gel electrophoretic separation of the chromosomes of strains from different lineage, obtained by recurrent mutagenesis and selection for improved cephalosporin C production (Smith *et al.*, 1991). Therefore, to obtain a maximum number of recombinants between production strains, the number of translocation for the strains involved should be low, indicating that 'sister' crosses are more advisable.

The role of protoplast fusion remains limited to the efficient production of heterokaryons and does not influence the subsequent genetic processes of diploidization and haploidization. The advantage of protoplast fusion is that heterokaryons can be produced at high frequency, which can reduce the introduction of selectable markers in the strains to be crossed to a minimum, allowing to use more easily, the parent strains with genetic markers that do not affect penicillin production, a method successfully applied at Gist Brocades in The Netherlands (Veenstra *et al.*, 1989).

Not only for strains used in the antibiotic industry, but also for other strains protoplast fusion experiments have been carried out for strain improvement. In *Penicillium caseicolum*, a species used in the dairy industry, protoplast fusion has been used to create strains with novel properties, which could be applied in the production of new dairy products. After fusion and selection, recombinant strains with changed morphological, lipolytical and proteolytical properties could be isolated and an anti-mucor property was transferred to all fusants, probably because of its mitochondrial location (Raymond *et al.*, 1986).

Less related species showed a more extensive incompatibility as concluded from heterokaryon morphology. All viable heterokaryons obtained between less related species consisted of colonies with irregular morphology and they produced only one type of parental spores. Nuclear fusion also occurred, but it gave rise to different types of aneuploid-prototrophic complementing parents also observed in the sporulating heterokaryon. The genetic background of these hybrids is not known, but it might now be identified by electrophoretic techniques as pulse field gel electrophoresis. The chromosomes of one species are preferentially lost, as they occur in interspecies hybrids of higher organisms, both plants and animals. Crosses carried out between distant species did not give viable progeny.

4.12 PROTOPLAST FUSION IN *ASPERGILLUS*

Industrially very important among the *Aspergillus* sp. is *A. niger*, mainly as source for the production of citric acid. Parasexual recombination in strain improvement programmes to obtain strains with enhanced citric acid production and improved substrate-use efficiency has already been tried since the detection of the parasexual cycle (Azevedo and Bonatelli, 1982). This approach remained limited to closely related strains due to difficulties in obtaining heterokaryons by hyphal anastomosis. By protoplast fusion, however, heterokaryons could readily be obtained between *A. niger* strains of divergent origin. Following crossing, diploid strains with increased citric acid production in the solid phase compared to the parents could be obtained, and haploid recombinants from the higher producing diploids showed further increase in critric acid production in both solid and sumberged cultures (Kirimura *et al.*, 1988; Martinkova *et al.*, 1990), indicating the usefulness of parasexual hybridization for strain improvement in this species. Similar results were obtained for the increase of α-amylase production in *A. niger*, for which recombinants were isolated with improved enzymatic activity. Das *et al.* (1989) have focussed protoplast fusion and genetic recombination in intra and inter strain crossing in *A. niger*. *A. niger* is a well known industrial fungus and is the source of several major products including organic acids and enzymes. Protoplast fusion crosses were carried out between sister strains and between divergent strains. The fusants were shown to be heterokaryons with enhanced activity.

Much work concerning interspecies protoplast fusion has been carried out with genetically well characterized *A. nidulans* (Peberdy, 1989). *A. nidulans* was crossed with several species classified in the *A. nidulans* group including *A. rugulosus, A. nidulans, A. echinulatus, A. quadrilineatus* and *A. violaceus*. All pairwise combinations between the five species were possible, except for the combination of *A. quadrilineatus* with *A. violaceus* and with *A. nidulans* var. *echinulatus*, but the latter species could be crossed with *A. quadrilineatus*. Three other species of the *A. nidulans* group, i.e. *A. stellatus, A. unguis* and *A. heterothallicus* did not produce viable interspecies heterokaryons. Crossing with *A. nidulans* containing well defined markers on each linkage group is helpful to examine the genetic relationship between the species and to investigate the fate of the genetic markers in the interspecies hybrids.

On analysis of hybrid segregants, it appeared that the species which could be crossed form a close association, but it suggested differences in the linkage group. With respect to mitochondrial DNA (mtDNA) similarities, as already mentioned, species which can be crossed have an overall mtDNA restriction map that is similar for the majority of restriction sites, although large differences in size may exist and it was observed that there is a high probability that mtDNA recombination occurs spontaneously in interspecies diploids.

As in *Penicillium* interspecies hybrids, it was observed in *Aspergillus* that hybridization can have significance with regard to gene regulation or the modification of gene products as detected by isozyme analysis and metabolite production.

A third group of *Aspergillus* that has been intensively investigated are the koji molds widely used in food and spirit industry for fermentation of sake, miso and soysauce. Koji molds produce several hydrolytic enzymes such as proteases and glutaminase and they are citric acid and acidic amylase hyperproducers. For strain-breeding purposes several of these strains including *A. oryzae, A. soja, A awamori* var. *Kawachi,* and *A. usamii* var. *shirousamii* have been protoplasted and fused either to form intraspecies diploids and recombinant or interspecies allodiploids and hybrids (Ushijima *et al.*, 1990).

4.13 PROTOPLAST FUSION IN EDIBLE MUSHROOMS (ANNE, 1993)

The most important edible mushroom species are members of the genus *Agaricus, Pleurotus, Lentinus, Flammulina, Coprinus,* etc. Breeding by conventional genetic means in order to obtain, e.g., more virus-resistant types of variants with longer stalks has always been very difficult, because of the special character of sexual reproduction. Also mutagenesis is hardly used due to problems in obtaining mutants from dikaryotic mycelium and the difficulties in isolating homokaryons. Breeding work with these species is often limited to collecting wild specimens at a superior quality or selection of a spontaneous mutant with improved properties for cultivation. However, in recent years much attention has been paid to the genetic improvement of edible mushrooms using protoplast fusion. *Pleurotus ostreatus* has been crossed with several other compatible or incompatible mating types including *P. columbinus*, *P. sajor-caju*, *P. florida* and *P. cornu-copiae*.

Interspecies crosses between compatible mating types resulting in fusion progeny of morphologically different variants and clamp connection and fruit body formation were observed. Many variants showed high fruit body formation. Interspecific fusion products between incompatible strains showed no true clamp connection and there was no fruit body formation, a phenomenon also observed when crossing *Coprinus machrorhizus* mutants as identical mating type (Kiguchi and Yanagi, 1985). These results indicate that although fusion and the stability of the progeny of fused protoplasts are not affected by mating characters, clamp connections and fruit bodies were formed mainly on the mycelium of compatible fusion products and not on the fusion progeny of incompatible species. Nevertheless, in some instances, e.g. in the fused progeny of the incompatible *P. ostreatus* and *P. eryngii* and of *P. florida* and *P. spodoleucus* fruit body formation was observed indicating the possibility of improving strains by fusing incompatible species (Go *et al.,* 1989). However, hybrids of incompatible strains would give rise to fruit body formation cannot be foreseen and can only be determined experimentally.

Concerning intergeneric crosses, a similar phenomenon as for other viable crosses in filamentous fungi occurred. In the hybrids one species was dominant over the other (Yoo, 1989), even after transfer of isolated nuclei into the protoplasts (Yoo *et al.,* 1987) instead of carrying out fusion between whole protoplasts.

4.14 PROTOPLAST FUSION IN *CEPHALOSPORIUM ACREMONIUM*

Although no specific examples dealing with the use of the protoplast fusion technique to improve the performance of biocontrol fungi could be found in the literature, several published examples dealing with other classes of fungi can be used to illustrate the usefulness of the possible applicability of the technique to genetically manipulate filamentous fungi. Hamlyn and Ball (1979) found that a strain of *Cephalosporium acremonium* that synthesizes the antibiotic cephalosporin C from inorganic sulfur yielded three times the amount of the antibiotic made by a distantly related strain. The former grew slowly without producing spores and the latter grew at about four times the rate of the other strain and sporulated abundantly. From crosses initiated by protoplast fusion made between these two strains, a prototroph was selected whose segregated progeny differed in growth rates and morphological characteristics. One segregant gave 40% more cephalosporin C than the higher titer prototrophic or genetically labeled parent and also grew faster and sporulated better than the prototroph.

4.15 PROTOPLAST FUSION IN YEASTS

Yeasts are fungi that have in common a predominantly unicellular vegetative stage. Their life cycle consists of a haploid and diploid or dikaryotic phase of which one of both phases is prevailing depending on the species. Sexually reproducing yeasts have in general a bipolar mating system. Mating is initiated by the agglutination of the haploid cells of different mating types following karyogamy gives rise to a zygote. From these diploid cells haploid ascospores (or sporidia in case of *Rhodosporidium*) are produced through meiosis during which recombination occurs because of the occurrence of meiotic recombination sexually reproducing yeast cells. Classical techniques of genetic breeding are potentially applicable in several industrially important yeasts. However, the majority of the industrial strains are polyploid or aneuploid, have aberrations in mating behaviour, poor sporulation or spore viability. Hence, conventional hybridization with these strains is difficult or impossible. Because of these limitations, protoplast fusion has been intensively used as an alternative to sexual breeding. Several reports have already appeared on yeast protoplast fusion, either chemically induced or by electrofusion and both intra and interspecies fusions have been successfully carried out (Spencer and Spencer, 1983; Morgan, 1983; Peberdy and Ferenczy, 1985; Ingolia and Wood, 1986). Protoplast fusion has been used extensively for the improvement of industrial strains of yeasts which are generally polyploid and not easily amenable to approaches such as sexual hybridization, mutagenesis and recombinant technology (Russell and Stewart, 1979).

Intraspecies fusions have been mentioned mainly for *Saccharomyces cerevisiae* but also for *Saccharomycopsis lipolytica, Saccharomyces diastaticus, Kluyveromyces lactis, Pichia guillermondii, S. pombe, Rhodosporidium toruloides, Candida utilis, C. maltosa, Apiotrichum curvatum, Lodderomyces elongisporus, Hansenula wingei* and *Trichosoporon adeninovorans.* Fusion occurred regardless of their mating type. Not only haploids of identical mating type could be fused but also haploids with diploids or diploids with diploids.

As a result of protoplast fusion, heterokaryons appeared or unstable diploids or stable diploids, but triploids and tetraploids as a consequence of multiple fusion bodies, have also been mentioned. Except for *Saccharomycopsis lipolytica,* protoplast fusion between heterothallic strains of like mating-type resulted in the formation of sporulaltion deficient hybrids, but when the diploids homozygous for the mating type locus were crossed or fused with haploid or diploid cells or protoplasts of opposite mating type, they developed spores. The observations indicated that mating type alleles in most instances not only control the initial step of mating, i.e. cell recognition and agglutination, but also meiosis and ascospore formation.

Diploids can segregate mitotically giving rise to haploid recombinatns, when treated with haploidizing agents such as p-fluorophenylalanine or methylbenzimidazol-2-yl carbamate. Spontaneous segregation to parental and recombinant types has also been observed following protoplast fusion, even in asexual species, but in the latter case karyogamy did not readily occur, in contrast to yeast from perfect genera for which the heterokaryotic stage is transient.

Following protoplast fusion, mitotic or meiotic recombination between genetically different nuclei occurs and therefore fusion offers the possibility of strain improvement for those strains in which sexual conjugation and genetic recombination are not observed due to the lack of a mating system or as a consequence of polyploidy and aneuploidy or sterility. In addition, cytoplasmic fusion also takes place by protoplast fusion. It is known that some characteristics of industrial yeasts are controlled by genes located on the mitochondrial genomes, e.g. uptake of some sugars, starch utilization, flocculation, flavour determinants and yeast killer factor. Not only do they contribute to strain improvement, but crosses involving petite mutants can help to elucidate or locate more precisely such functions on the mitochondria. In addition, for reasons mentioned earlier petite mutants are often used in crosses aimed at strain improvement.

a. Protoplast Fusion for Improving Brewery and Distillery Yeast (Anne, 1993)

Saccharomyces diastaticus has been widely used to improve brewery and distillery yeasts, because of its glucoamylase activity enabling the yeast to utilize dextrin and to produce low carbohydrate beer (Janderova *et al.,* 1990). In this manner by fusing *S. uvarum* and *S. diastaticus* a brewery yeast strain was isolated which can utilize wort carbohydrates that normally are unfermentable. *S. diastaticus* has the disadvantage that it carries a phenolic off-flavour (*pof*) gene which can cause phenolic off-flavors in beer, making it unpalatable but by using mutants *pof* or by backcrossing and selection for suitable recombinants this problem can be overcome. An interesting finding from protoplast fusion with brewery yeasts which usually sporulate poorly was also that the fusion products had a greatly increased degree of sporulation and showed normal asci formation. As a result, genetic analysis of polyploid parents is rendered possible. Wine yeasts have been improved by using protoplast fusion with respect to fermentation performance and phenological properties (Yokomori *et al.,* 1989).

b. Construction of a Flocculant Yeast with Killer Character (Javadekar *et al.,* 1995)

A killer toxin producing flocculant yeast strain was constructed by protoplast fusion between a highly flocculent strain and a killer strain of *S. cerevisiae.* The loss of flocculance in molasses of regenerated fusants remained unexplained. Protoplast fusion has been used extensively for the improvement of industrial strains of yeast which are generally polyploid and not easily amenable to approaches such as sexual hybridization, mutagenesis and recombinant DNA technology.

A desirable property to control killer sensitive wild yeasts during industrial fermentation particularly in continuous operations over prolonged periods in the killer character (Vondrej, 1987). The transfer of the killer character to laboratory and industrial strains has been achieved in many laboratories by various methods (Hara *et al.,* 1981), bred killer strains of *S. cerevisiae* by back crossing. Construction of killer strain by gene replacement was described by Boone *et al.* (1990). Salek *et al.* (1992) used electrotransformation to get stable hybrids having killer activity. Flocculation is an important prerequisite in yeasts used in industrial fermentation and construction of flocculant strains also has been carried out by protoplast fusion (De Figueroa *et al.,* 1984 and Wateri *et al.,* 1990) as well as by electrofusion (Urano *et al.,* 1990; 1993). Javadekar *et al.,* 1995 have reported the construction of a killer toxin producing flocculant yeast strain by protoplast fusion between a highly floccu-lant strain and a killer strain.

4.16 INTERGENERIC PROTOPLAST FUSION BETWEEN *ASPERGILLUS NIGER* AND *TRICHODERMA VIRIDE*

Recently, intra and interspecific protoplast fusion have been carried out in various filamentous fungi in order to obtain heterokaryons, heterodiploids and recombinants. On the other hand, intergeneric protoplast fusion has been carried out between *Cephalosporium acremonium* and *Penicillium chrysogenum* and *Aspergillus oryzae* and *Rhizopus javanicus,* In crosses between *C. acremonium* and *E. glabra,* and *A. oryzae* and *R. javanicus,* the fusants showed recombinant-type properties. In the cross between *C. acremonium* and *P. chrysogenum*, the fusants showed heterokaryon-like properties; however, the results were not presented in detail.

Two types, of fusants: heterokaryons showing mixed morphologies, between those of *A. niger* and *T. viride,* and presumable recombinants showing an *A. niger* type morphology.

Auxotrophic mutant strains Y-(b) (nicotinic acid-requiring, brown conidia), from *A. niger* Yang no. 2, and *T. viride* M5S51 (leucine-requiring), from *T. viride* WU-36B, were used for protoplast fusion. These mutant strains were induced by UV irradiation, the reversion frequencies being less than 20 $\times 10^{-8}$ per conidium.

Fusants that appeared were picked up and then cultivated on MM plates. The fusion frequency was 10^{-5} to 10^{-4} resulting to 123 fusants, which are classifiable into two types according to their morphologies. As to the first type, 31 fusant strains formed colonies like those of *A. niger* Y-(b) on MM and SM. They grew as well as the prototrophic parental strain, *A. niger* Y-(b), whereas they were prototrophic. The appearance of this type of fusant strain did not change on subcultivation on MM and SM, moreover, they were stable on SM containing 1.25 μg/ml of benomyl as a haploidizing agent. From these results, these fusant strains were assumed to be recombinants. As to the second type, 92 fusant strains formed colonies showing mixed morphologies, between those of *A. niger* and *T. viride*. The fusant strains of the second type were judged to be heterokaryons and these strains formed conida showing the same nutritional requirements as those of *A. niger* Y-(b) and *T. viride* M5S51 they grew more slowly than the protoplasts parental strains, *A. niger* Yang no. 2 and *T. viride* WU-36B. When any part of the mycelia was picked up and subcultivated on MM, it formed a colony which showed the same morphology as that of the original fusant strain. In addition, the morphology of the second type of fusant strains was completely different from that if the prototrophic parents strains. *A. niger* Yang no.2 and *T. viride* WU-36B, were cultivated together on MM. When conidia of *A. niger* Y-(b) and *T. viride* M5S51 were mixed and inoculated onto MM, no colony appeared. When the mycelial mat of a fusant colony on MM was picked up and transferred onto SM the resulting colony showed a morphology like that of either *A. niger* Y-(b) or *T. viride* M5S51.

Conclusion

The fungal protoplast fusion and strain improvement is of global interest as the applications and utility value of fungi and fungal products are tremendous that they are focussed in every field of science. Their products range from low molecular weight primary metabolites to complex secondary metabolites including antibiotics and alkaloids. In addition, vitamins and many enzymes are produced by fungi. Fungi play a key role in food industry. Yeasts are indispensable for alcoholic fermentation, besides, fungi are also useful in baking and cheese making. Mushroom fungi are edible ones and are used for direct consumption. Above all filamentous fungi are serious pathogens and also biocontrol agents. Wealth of the above information about filamentous fungi distinctly expose the importance of strain improvement for enhanced or over expression of products, integration of characters and indepth probe to understand the pathogens. History of strain improvement shows mutation and selection as methods of strain improvement. But there are fungi with no sexual reproduction but with tremendous input in industrial application. Even fungi with sexual cycle show several problems of homothallism, anueploidy or triploidy, aberrations in mating behaviour, poor sporulation and spore viability which interupts the classical strain improvement. In view of several obstacles encountered in parasexual cycle development of protoplast fusion technique was putforth by several scientists. This technique overrules incompatibility and is applicable for strains of the same species and also different species. Though protoplast fusion technique was easy to manipulate and eventhough more reports appeared for fungi, yet it is considered new because the protocol varies with fungi and observations made over several fungi give varied results. Though several reviews have been published (Potrykus *et al.*, 1983; Anne, 1985; Peberdy and Ferenczy, 1985; Peberdy, 1989) regarding protoplast fusion, the present book gives a collective information based on the author and her school's personal contributions in the application of protoplast and protoplast fusion technology to understand plant pathogens and biocontrol agents.

This book is a ready reconer for the researchers and scientists who are involved in strain improvement technology. Collection of informations are given in order to avoid disappointment in this technique. Observations on different filamentous fungi and the combined presentation in terms of plates and figures will be useful for the beginners to look at the variations in the protoplast release, germination and elongation of primary hypha. Literature cited are the practical evidences for the researchers. Factors affecting the release of protoplasts from different filamentous and yeast fungi are records of informations and based on definite theory of understanding using protoplasts. Viable protplasts for wall biogenesis, cell free enzymes and for the isolation of organelle and organellar DNA (mitochondrial DNA) have been well documented with the author's work. Informations pooled in regarding regeneration medium, regeneration types, factors affecting regeneration and various morphology of regenerating hyphae will be useful for the researchers to proceeds without hesitation of error with any type of filamentous fungi for strain improvement. Regeneration process of fungal protoplast is of prime importance to clarify protoplasts a viable one. Anucleated and nucleated protoplasts are compared critically to assess the importance of chromosomal DNA in the cell wall biogenesis of filamentous and yeast fungi. Classifications of binucleate, trinucleate and multinucleate conditions in the present book is a speculation based on the number of regenerating hyphae observed by the author in filamentous fungi. Exploitation of individual protoplast regenerated colonies for natural selection of virulent and avirulent clone (plant pathogen) potential clone (enzyme producer) drug or fungicide sensitive and resistant clone has been confirmed and assessed by the author of this book Regeneration of protoplasts is important for successful protoplast fusion. Regeneration occurs rapidly within hours provided suitable conditions are optimized. Regeneration range from 48h to 1 week depending on the type of fungi. In this book extensive importance is given for regeneration types and factors affecting regeneration. All most all reports of filamentous fungi are presented here with reference to their cell wall biogenesis and regeneration. Protoplast fusion technology is also covered in depth to have a clear vision on the strategies for selecting heterokaryons and conditions for successful fusion. Of course the mechanism of membrane fusion reasoning out the usefulness, simplicity and authenticity of this technique have been extensively covered. Application of this technology will be increasing in the next millenium and the present book will serve as a dictionary to all those who are interested in strain improvement of fungi. As among the wide variety of strain improvement techniques, protoplast fusion seems to be an efficient way to induce genetic recombination in fungi, which have n sexual cycle.

Bibliography

Abe M, Umetsu H, Nakai T and Sasage P. 1982. Regeneration and fusion of mycelial protoplasts of *Trichoderma matsutake*. *Agri. Biol. Chem.* 46: 1955-1957.

Adams G, Johnson N, Leslie JF. and Hart LP. 1987. Heterokaryons of *Gibberella zeae* formed following hyphal anastomosis or protoplast fusion. *Exp. Mycol.* 11: 339-353.

Ahkong QF, Cramp FC, Fisher D, Howell JI, Tampion W, Verrinder M and Lucy JA. 1973. Chemically-induced and thermally induced cell fusion: lipid-lipid interactions. *Nature* 242: 215-217.

Akamatsu T and Sekiguchi J. 1981. Studies on regeneration media of *Bacillus subtilis* protoplasts. *Agri. Biol. Chem.* 45: 2887-2894.

Akamatsu K, Kamada T and Takemaru T. 1983. Release and regeneration of protoplasts from the oidia of *Coprinus cinereus*. *Trans. Mycol. Soc.* Japan 24: 173-184.

Alexopoulous CJ and Mims CW. 1993. *Introductory Mycology*. Wiley Eastern Ltd. Publishers.

Algranati ID, Carminalli H and Cabid E. 1963. *Biochim. Biophy. Res. Commun.* 12: 504-509.

Allcock ER, Reid SJ, Jones DT and Woods DR. 1982. *Clostridium acetobutylicum* protoplast formation and regeneration. *Appl. Environ. Microbiol.* 43: 719-721.

Allmark BM, Morgan AJ and Whitaker PA. 1978. *Mol. Gen. Genet.* 159: 297-299.

Anderson FB and Millbank JW. 1966. Protoplast formation and yeast cell wall structure, the action of the enzyme of the snail helix pomatia. *Biochem.* J. 99: 682-687.

Annamalai P. 1989. Molecular basis of edipehnphos resistance in *Drechlera oryzae* (Breda De Haan) Subram and jain and control measures. Ph.D. Thesis. University of Madras, Madras.

Annamalai P and Lalithakumari D. 1991. Isolation and regeneration of protoplasts from mycelium of *Drechslera oryzae. Journal Plant Disease and Protection* 98: 197-204.

Annamalai P and Lalithakumari D. 1993. Biochemical changes and mechanism of resistance in *B.oryzae* to ediphenphos. *J. Plant. Dis. Prot.* 100: 497-507.

Anne J. 1977. Somatic hybridization between *Penicillium* species after induced fusion of their protoplasts. *Agricultura* 25: 1-117.

Anne J, Eyssen H and De Somer P. 1974. Formation and regeneration of *Penicillium chrysogenum* protoplasts. *Arch. of Microbiol.* 98: 159-166.

Anne J and Peberdy JF. 1975. Conditions for induced fusion of fungal protoplasts in polyethylene glycol solutions. *Arch. of Microbiol.* 105: 201-205.

Anne J and Peberdy JF. 1976. Induced fusion of fungal protoplasts following treatment with polyethylene glycol. *J. Gen. Microbiol.* 92: 413-417.

Anne J, Eyssen H and De Somer, P. 1976. Somatic hybridization of *Penicillium roquefortii* with *P. chrysogenum* after protoplast fusion. *Nature* 262: 719-721.

Anne J and Eyssen H. 1978. Isolation of interspecies hybrids of *Penicillium citrinum* and *Penicillium cyanofulvum* following protoplast fusion. *FEMS Microbial. Lett.* 4: 87-90.

Anne J, Van Mellaert L and Eyssen H. 1990. Optimum conditions for efficient transformation of *Streptomyces venezuelae* protoplasts. *Appl. Microbiol. Biotechnol.* 32: 431-435.

Anne J. 1993. Cell Fusion In: Biotechnology, Vol. 2, Rehm, HJ and Reed G (eds), VCH, Weinheim, pp. 93-139.

Antonov PA. 1990. Thermofusion of cells. *Biochem. Biophys. Acta.* 1051: 279-281.

Arima K and Takano I. 1979. Multiple fusion of protoplasts in *Saccharomyces* yeasts. *Trans. Mycol. Soc.* Japan 19: 181.

Asai T, Okuno T and Matsuura K. 1986. Isolation and germination type of protoplasts of spores and hyphae of *Pyricularia oryzae. Ann. Phytopath. Soc.* Japan 52: 843-849.

Auer D, Brandner G and Bodemer W. 1976. Dielectric breakdown of the red blood cell membrane and uptake of SV40 DNA and mammalian cell RNA. *Naturwissenschaften* 63: 391.

Azevado JL and Bonatelli (Jr) R. 1982. Genetics and over production of organic acids. In: Over production of Microbial products, (Krumphanzl V, Sikyta B and Vanek Z (Eds), Academic Press, London, pp. 439-450.

Bachmann, BJ and Bonner DM. 1959. Protoplasts from *Neurospora crassa. J. Bact.* 78: 550-556.

Bacon JSD, Jones D and Ottolenghi P. 1969. *J. Bact.* 99: 885-887.

Ball C. 1984 b Filamentous fungi. In: Genetics and breeding of industrial microorganisms, Ball C (Ed.), CRC, Boca Raton. FL. 159-188.

Bartnicki-Garcia and Lippman, E. 1966. Liberation of protoplast from mycelium of Phytophthora Journal of General Microbiology 42, 411-416.

Bartnicki-Garcia S. 1973. Fundamental aspects of hyphal morphogenesis. *Symp. Soc. Gen. Microbiol.* 23: 245-267.

Basett RA, Chain EB and Corbett K. 1973. Biosynthesis of ergatomine by *Claviceps purpurea. App. Microbiol.* 134: 1-10.

Becher D and Bottcher F. Hybridization of *Rhodosporidium toruloides* by protoplast fusion. (This volume).

Beggs JD. 1978. Transformation of yeast by replicating hybrid plasmid. *Nature* 275: 104-109.

Benitez T, Ramos S and Garcia Acha I. 1975. Protoplasts of *Trichoderma viride*: Formation and regeneration. *Arch. Microbiol.* 103: 191-204.

Berkuber and Reca, 1969.

Berliner MD, Carbonell LM and Biundo N. 1972. *Mycologia* 64: 708-721.

Berliner MK and Reca ME. 1969. *Mycopathol. Mycol. Appl.* 37: 81-85.

Berliner, MK and Reca ME. 1971. Studies on protoplast induction in the yeast phase of *Histoplasma capsulatum* by magnesium sulfate and 2-deoxy-D-glucose. *Mycologia* 6: 1164-1172.

Binding H and Weber HJ. 1974. The isolation, regeneration and fusion of *Phycomyces* protoplasts. *Mol. Gen. Genet.* 135: 273-276.

Birnboim HC. 1971. New method for extraction of ribonucleic acid and polyribosomes from *Schizosaccharomyces pombe. J. Bact.* 107: 659-663.

Blumenthal R. 1987. Membrane fusion. *Curr. Top. Membr. Transp.* 29: 203-254.

Boizet B, Flickinger JL and Chassy BM. 1988. Transfection of *Lactobacillus bulgaricus* protoplasts by bacteriophage DNA. *Appl. Environ. Microbiol* 54: 3014-3018.

Boland GJ and Smith EA. 1991. Variation in cultural morphology and virulence among protoplast-regenerated isolates of *Sclerotinia sclerotiorum. Phytopathol.* 81: 766-770.

Boni LT, Hah JV, Hui SV, Mukherjee P, Ho JT and Jung CY. 1984. Aggregation and fusion of unilamellar vesicles by polyethylene glycol. *Biochem. Biophys. Acta.* 775: 409-418.

Boone C, Sdieu AM, Wagner J, Degre R, Sanchez C and Bussey H. 1990. Integration of the yeast KI killer toxin gene into the genome of marked wine yeasts and its effects on vinification. *Am. J. Enol. Vitic.* 41: 37-42.

Bottcher D, Becher U, Klinner, Samsonova IA and Schilova B. Genetic structure of yeast hybrids constructed by protoplast fusion. (This volume)

Boultan A.A. 1965 Exptl. Coll Res 37: 343

Bradshaw RE, Kang-Up Lee and Peberdy JF. 1983. Aspects of genetic interaction in hybrids of *Aspergillus nidulans* and *Aspergillus rugulosus* obtained by protoplast fusion. *J. Gen. Microbiol.* 129: 3525-3533.

Bradley SG. 1959. Protoplasts of *Streptomyces griseus* and *Nocardia paraguayensis. J. Bact.* 77: 115-116.

Braun PJ and Heisler A. 1990. Isolation and cell wall regeneration of protoplasts from Botrytis *cinerea. Pers. J. Phytopathol.* 128: 293-298.

Brown JP. 1971. Susceptibility of the cell walls of some yeasts to lysis by enzymes of *Helix pomatia. Can. J. Microbiol.* 17: 205-208.

Burger K. 1991. The mechanism of influenza virus membrane fusion. Ph.D. Thesis, University of Utrecht, The Netherlands.

Caporale LH, Chartzam N, Tocci M and DeHaven. 1990. Protoplast fusion in microtiter plates for expression cloning in mammalian cell: demonstration of feasibility using membrane-bound alkaline phosphatase as a reporter enzyme. *Gene* 87: 285-289.

Caso JL, Hardisson C and Suarez JE. 1987. Transfection in *Micromonospora* spp. *Appl. Environ. Microbiol.* 53: 2544-2547.

Chakraborty, BN and Kapoor M. 1990. Transformation of filamentous fungi by electroporation. *Nucleic Acids Res.* 18: 6737.

Chand PK, Davey MR, Power JB and Cocking EC. 1988. An improved procedure for protoplast fusion induced by electric fields. *Cell Biophys.* 14: 231-243.

Chang LT and Terry CA. 1973. Intergeneric complementation of glucoamylase and citric acid production in two species of *Aspergillus. Appl. Microbiol.* 25: 890-895.

Chang LT, Terasaka T and Elander RP. 1982. Protoplast fusion in industrial fungi. *Dev. Ind. Microbiol.* 23: 21-29.

Chang St, Li GSFD and Peberdy JF. 1985. Isolation of protoplast from edible fungi. *Mircen Journal* 1: 185-194.

Chang DC, Hunt JR and Gao PQ. 1989. Effects of pH on cell fusion induced by electric fields. *Cell Biophys.* 14: 231-243.

Chapel M, Montane MH, Ranty B, Teissie J and Alibert G. 1986. Viable somatic hybrids are obtained by direct current electrofusion of chemically aggregated plant protoplasts. *FEBS Lett.* 196: 79-86.

Chellappa G. 1988. Strain improvement through protoplast fusion technology for enhanced lytic enzyme production. Ph.D. Thesis, University of Madras, India.

Chen Z, Wojcik SF and Welder NE. 1986. Genetic analysis of *Bacillus stearothermophilus* by protoplast fusion. *J. Bacteriol.* 165: 994-1001.

Choi SH, Byong KK, Kim HW, Choi EC, Kim YC and Park YB. 1987. Studies on Protpolast Formation and Regeneration on *Ganoderma lucidum. Arch. Pharm. Res.* 10: 158-164.

Cirillo. VP. 1966. In In Symposium uber Hefe-Protoplasten, Jena. Ed. R. Muller, Academic-Verlag, Berlin.

Cocking EC. 1984. Plant cell fusion: Transformations using plant and bacterial protoplasts. In: Cell fusion: Gene transfer and transformation, Beers (Jr) RF and Bassett EJ, (Eds), Raven Press, New York, pp. 139-144.

Cole GT and Aldrich HC. 1971. Ultrastructure of conidiogenesis in *Scopulariapsis brevicaulis. Can. J. Bot.* 49: 745-749.

Constabel F and Kao KN. 1974. Agglutination and fusion of plant protoplasts by polyethylene glycol. *Can. J. Bot.* 52: 1603-1606.

Corner TR and Marquis RE. 1969. *Biochim. Biophy. Acta.* 183: 544-558.

Cramp FC and Lucy JA. 1974. Glyceryl mono oleate as a fusogen for the formation of heterokaryons and interspecific hybrids. *Exp. Cell. Res.* 87: 107-110.

Croce CM, Sawicki W, Kritchevsky D and Koprowski, H. 1971. Induction of homokaryocyte, heterokaryocyte and hybrid formation by Lysolecithin. *Exp. Cell. Res.* 67: 427-435.

Croft JH. 1985. Protoplast fusion and incompatibility in *Aspergillus.* In: Fungal protoplasts: Application in biochemistry and genetics, Peberdy JF and Ferenczy L (Eds), Marcel Dekker Inc., New York, USA, pp. 225-240.

Dales RBG and Croft JH. 1977. Protoplast fusion and the isolation of heterokaryons and diploids from vegetatively incompatible strains of *Aspergillus nedulans FEMS Microbiol. Lett.* 1: 201-204.

Darkan, MA. 1962. Title of the article. *Appl. Microbiol.* 10: 387-393.

Darling S, Theilade J and Birch-Andersen A. 1969. Kinetic and morphological observations on *Saccharomyces cerevisiae* during spheroplast formation. *J. Bact.* 98: 797-810.

Das A, Gokhale DV and Peberdy JF. 1989. Protoplast fusion and genetic recombination in intrastrain and interstrain crossiing in *Aspergillus niger. Enzyme Microb. Technol.* 11: 2-5.

Davis B. 1985. Factors influencing protoplast isolation. In: Fungal Protoplasts: Applications in Biochemistry and Genetics, Peberdy JF and Ferenczy L (Eds), Marcel Dekker, New York, pp. 45-72.

De Figueroa LI, de Richard NF and de van Broock MR. 1984. Use of somatic fusion method to introduce the flocculation property into *Saccharomyces distaticus. Biotechnol. Lett.* 6: 587-592.

De Kloet SR, Van Wermeskerrken RKA and Koningsberger VV. 1961. *Biochim. Biophy. Acta* 47: 138-148.

Deb JK, Malik S, Ghosh VK, Mathai S and Sethi R. 1990. Intergeneric protoplast fusion between xylanase producing—*Bacillus subtilis* LYT and *Corynebacterium acetoacidophilus* ATCC 21476. *FEMS Microbiol. Lett.* 71: 2887-2892.

Delgado JM and Herrerra LS. 1981. protoplast fusion in the yeast *Candida utilis. Acta Microbiol. Acad.* Sci. Hungary? 28: 339-345.

Delorme D. 1989. Transformation of *Saccharomyces cerevisiae* by electroporation. *Appl. Environ. Microbiol.* 55: 2242-2246.

Deutch CE and Parry JM. 1974. Sphaeroplast formation in yeast during the transition from exponential phase to stationary phase. *J. Gen. Microbiol.* 80: 259-268.

De Vries OMH and Wessels JGH. 1972. Release of protoplasts from *Schizophyllum commune* by a lytic enzyme preparation from *Tricho-derma viride. J. Gen. Microbiol.* 76: 319-330.

De Vries OMH and Wessels JGH. 1975. Chemical analysis of cell wall regeneration and reversion of protoplasts from *Schizophyllum commune. Arch. Microbiol.* 102: 209-218.

De Waard MA. 1976. Formation of protoplasts from *Ustilago maydis. Antonie van Leeuwenhoek* 42: 211-216.

Dooijewaard-Kloosterziel AMP, Sietsma JH and Wouters JTM. 1973. Formation and regeneration of *Geotrichum candidum* protoplasts. *J. Gen. Microbiol.* 74: 205-209.

Douglas RJ, Robinson JB and Corke CT. 1958. On the formation of protoplast like structures from *Streptomyces. Can. J. Microbiol* 4: 551-554.

Duercksen JD. 1964. *Biochim. Biophy. Acta.* 87: 123-140.

Dziengel A, Held W, Schlanderer G and Dellweg H. 1977. *Eur. J. Appl. Microbiol.* 4: 21-27.

Earl AJ, Turner G, Croft JH, Dales RBG, Lazarus CM, Lundsorf H and Kuntzel H. 1981. High frequency transfer of species-specific mitochondrial DNA sequences between members of Aspergillaceae. *Curr. Genet.* 3: 221-228.

Eddy AA and Williamson DH. 1957. A method for isolating protoplasts from yeast. *Nature* 179: 1252-1253.

Eddy AA and Williamson DH. 1959. Formation of aberrant cell walls and spores by the growing yeast protoplast. *Nature* 183: 1101-1104.

Editmann A and Schauz K. 1992. Cryopreservation of protoplasts from sporidia of *Ustilago maydis. Mycol. Res.* 98: 318-320.

Elavarasan A. 1996. Genetics of the stability and survival of fitness of *Rhizoctonia solani* (Kuhn) laboratory mutants resistant to benzimidazole and dicarboximide fungicides, Ph.D. Thesis, University of Madras, Madras, India.

Elorza MW, Arst HN, Cove DJ and Scazzochio C. 1969. *J. Bact.* 99: 113-115.

Emerson S and Emerson MR. 1958. Production, regeneration and reversion of protoplast-like structures in the osmotic strain of *Neurospora crassa. Proc. Nat. Acad.* Sci. USA 44: 668-671.

Eveleigh DE, Sietsma JH and Haskins RH. 1968. The involvement of cellulase and laminarinase in the formation of *Pythium* protoplasts. *J. Gen. Microbiol.* 52: 89-97.

Fantini AA. 1962. Genetics and antibiotic production of *Emericellopsis* species. *Genetics* 47: 161-177.

Fawcett PA, Loder PB, Duncan MJ, Beesley, TJ and Abraham EP. 1973. Formation and properties of protoplasts from antibiotic-producing strains of *Penicillium chrysogenum* and *Cephalosporium acremonium. J. Gen. Microbiol.* 79: 293-309.

Felgner PL, Gadek TR, Holm M, Roman R, Chan HW, Wenz, M, Northrop JP Ringold GM and Danielsen M. 1987. Lipofection: a highly efficient, lipid-mediated DNA transfection procedure. *Proc. Nat. Acad. Sci* USA 84: 7413-7417.

Ferenczy L, Kevei F and Zsolt J. 1974. Fusion of fungal protoplasts. *Nature* 248: 793-794.

Ferenczy L, Kevei F and Szegedi M. 1975a. Increased fusion frequency of *Aspergillus nidulans* protoplasts. Experientia, 31: 50-52.

Ferenczy L, Kevei F and Zsolt J. 1975b. High frequency fusion of protoplasts. Experentia. 31: 1028-1029.

Ferenczy L. 1976. Some characteristics of intra and inter specific protoplast fusion products of *Aspergillus nidulans* and *Aspergillus fumigatus*. In: Cell Genetics in higher plants Dudits D:, Farkas GL and Maliga P (Eds), Akademiai Kiado, Budapest, pp. 171.

Ferenczy L. 1976. In: Cell genetics in higher plants, Dudits D, Farkas GL and Maliga P (Eds), Budapest Akademiai Kiado, Budapest, pp. 171-182.

Ferenczy L, Kevei F and Szegedi M. 1976. Fusion of fungal protoplast induced by polyethylene glycol. In: Microbial and Plant Protoplasts, Peberdy JF, Rose AH, Rogers HJ and Cocking EC (Eds), Academic Press, New York-London-San Francisco, pp. 177.

Ferenczy L, Kevei F and Szegedi M. 1976. Factors affecting high-frequency fungal protoplast fusion, Experientia 32: 1156.

Ferenczy L and Maraz A. 1977. Transfer of mitochondria by protoplast fusion in *Saccharomyces cerevisiaea. Nature* (London) 268: 524-525.

Ferenzy L and Pesti M. 1982. Transfer of isolated nuclei into protoplasts of *Saccharomyces cerevisiae. Curr. Microbiol.* 7:157-160.

Ferenczy L, Vallin C and Maraz A. 1979. A method of protoplast fusion for *Canadida tropicalis* and other yeasts. Proc. 6th Intern. Spec. Symp., Yeasts Keszthely, pp. 85.

Ferenczy L. 1981. Microbial protoplast fusion. In: Genetics as a tool in Microbiology, Glover SW and Hopwood DA (Eds), Cambridge Univrsity Press, Cambridge, pp. 1-34.

Ferenczy L. 1984. Fungal protoplast fusion: basic and applied aspects. In: Cell fusion: Gene transfer and transformation, Beer (Jr) RF and Bassett EG (Eds), Raven Press, New York, pp. 145-169.

Fleischer ER and Vary PS. 1985. Genetic analysis of fusion recombinants and presence of noncomplementing diploids in *Bacillus megaterium. J. Gen. Microbiol.* 131: 919-926.

Floss HG. 1976. Biosynthesis of ergot alkaloid and related compounds. *Tetrahedron* 32: 873-912.

Fournier P, Provost A, Bourguignon C and Heslot H. 1977. Recombination after protoplast fusion in the yeast *Candida tropicalis. Arch. Microbiol* 115: 143.

Foury F and Goffeau A. 1973. Combination of 2-Deoxy glucose and snail gut enzyme treatments for preparing sphaeroplasts of *Schizosaccharomyces pombe, J. Gen. Microbiol.* 75: 227-229.

Fodor K, Hadlaczky G and Alfodi L. 1975. Reversion of *Bacillus megaterium* protoplasts to the bacillary form. *J. Bacteriol.* 121: 390-391.

Friss J and Ottolenghi P. 1959. *C.r. Trav. Lab. Carlsberg.* 31:269-271.

Gabor MH and Hotchkiss RD. 1979. Parameters governing bacterial regeneration and genetic recombination after fusion of *Bacillus subtilis* protoplasts. *J. Bact.* 137: 1346-1353.

Garcia Acha I, Rodriguez-Aguirre J and Villanueva JR. 1963. *Microbiol. Espan.* 16: 141-148.

Garcia AI, Lopez-Belmonte F and Villanueva JR. 1966. Regeneration of mycelial protoplasts of *Fusarium culmorum. J. Gen. Microbiol.* 45: 515-523.

Garcia Mendoza C and Villanuea JR. 1967. *Biochim. Biophy. Acta.* 135: 189-197.

Gascon S, Ochoa AG and Villanueva JR. 1965. Production of yeast and mold protoplasts by treatment with the streptozyme of Micromonospora *Can. J. Microbiol.* 11: 573-580.

Gascon S and Villanueva JR. 1965. Magnesium sulphate as stabilizer during liberation of yeast and mould protoplasts. *Nature* 205: 822-823.

Gasson MJ. 1980. Production, regeneration and fusion of protoplasts in lactic streptococci. *FEMS Microbiol.* Lett. 9: 99-102.

Genther FJ and Borgia PT. 1978. Spheroplast fusion and heterokaryon formation in *Mucor racemosus J. Bacteriol.* 134: 349.

Gibson RK and Peberdy JF. 1972. Fine structure of protoplasts of *Aspergillus nidulans. J. Gen. Microbiol.* 72: 529-538.

Girbardt M. 1969. Die Ultrastruktur der Apikalregion von Pilzhyphen. *Protoplasma* 67: 413-441.

Go SJ Shin GC and Yoo YB. 1985. Protoplast formation, regneration and reversion in *Pleurotus ostretus* and *P. sajor-caju. Kor. J. Mycol.* 13: 167-177.

Go SJ, You CH and Shin GC. 1989. Effects of incompatibility on protoplast fusion between intra and interspecies in basidiomycete: *Pleurotus spp. Kor. J. Mycol.* 17: 137-144.

Gramp FC and Lucy JA. 1974. Glyceryl monoleate as a fusogen for the formation of heterokaryons and interspecific hybrids. *Exp. Cell Res.* 87: 107-110.

Grove SN. 1978: In: The filamentous fungi, Smith JE and Berry DR (Eds), London Edward Arnold, London 3: 28-50.

Grove SN and Bracker CE. 1970. Protoplasmic organization of hyphal tips among fungi: Vesicles and Spitzenkorper. *J. Bact.* 104: 989-1009.

Groves DP and Oliver SG. 1984. Formation of intergeneric hybrids of yeast by protoplast fusion of *Yarrowia* and *Kluyveromyces* species. *Curr. Genet.* 8: 49-55.

Griffin DH, Devit M and Tuori R. 1989. Protoplast formation and transformation of *Hypoxylon mammatum.* Phytopath. 79: 1204.

Guerra-Tschuschke I, Martin I and Gonzales MT. 1991. Polyethylene glycol induced internationalization of bacteria into fungal protoplasts: Electron microscopic study and optimization of experimental conditions. *Appl. Environ. Microbiol.* 57: 1516-1522.

Gunge N and Sakaguchi 1979. Fusion of mitochondria with protoplast in *Saccharomyces cerevisiea. Mol. Gen. Genet.* 170: 243-247.

Gunge N and Tamaru, A. 1978 and *Jpn. J. Genet.* 53: 41-59.

Gwinn KD and Daub ME. 1988. Regenerating protoplasts from *Cercospora* and *Neurospora* differ in their respone to cercosporin. *Phytopath.* 78: 414-418.

Hara S, Imura Y, Oyama H, Kozeki T, Kitano K and Otsuka K. 1981. The breeding of cryophilic killer wine yeasts. *Agric. Biol. Chem.* 45: 1327-1334.

Haigler C, Brown (Jr) RM and Benziman M. 1980. Calcoflor white ST alter the *in vivo* assembly of cellulose microfibrils. *Science* 210: 903-906.

Hammil TM. 1972. Electron microscopy of phialoconidiogenesis in *Metarrhizium anisopliae. Am. J. Bot.* 58: 88.

Hamilton JG and Calvet J. 1964. Production of protoplasts in an osmotic mutant of *Neurospora crassa* without added enzyme. *J. Bact.* 88: 1084.

Hamlyn PF and Ball C. 1979. Recombination studies with *Cephalosporium acremonium*. In: Genetics of Industrial Microorganisms, Sebek OK and Laskin AI (Eds). American Society of Microbiology, Washington DC, pp. 185-191.

Hamlyn PF, Bradshaw RE, Mellonma FM, Santiago CM, Wilson JM and Peberdy JF. 1981. Efficient protoplast isolation from fungi using commercial enzymes. *Enz. Microbiol. Technol.* 3: 321-325.

Harling R, Kenyon L, Lewis BG, Oliver RP, Turner JG and Coddington A. 1988. Conditions for efficient isolation and regeneration of protoplasts from *Fulvia fulva. J. Phytopathol.* 122: 143-146.

Harris GM. 1982. Protoplasts from *Gibberella fujikuroi. Phytopath.* 72: 849-853.

Harris H and Watkins JF. 1965. Hybrid cells derived from mouse and man: artificial heterokaryons of mammalian cells from different species. *Nature* 205: 640-646.

Hans, SS, Lee YY and Yu SH. 1988. Variation in the pathogenicity of protoplast regenerated isolates of *Pyricularia oryzae. Kor. J. Plant Path.* 4: 156-160.

Hasegawa S, Matsui C, Nagata T and Syono K. 1983. Cytological study of the introduction of *Agrobacterium tumefaciens* spheroplasts into *Vinca rosea* protoplasts. *Can. J. Bot.* 61: 1052-1057.

Hashiba T and Yamada M. 1982. Formation and purification of protoplasts from *Rhizoctonia solani. Phytopath.* 72: 849-853.

Hashiba T and Yamada M. 1984. Intraspecific protoplast fusion between auxotrophic mutants of *Rhizoctonia solani. Phytopath.* 74: 398-401.

Hastie AC. 1981. The genetics of conidial fungi. In: Biology of conidial fungi, Vol. 2 : GT Cole and Kendricks B (Eds), Academic Press, New York. pp. 511-547.

Havelkova M. 1969. *Fol. Microbiol.* 14: 155-164.

Hawker LE, Thomas B and Beckett A. 1970. An electron microscopic study of the structure and germination of conidia of *Cunninghamella elegans* (Lender) *J. Gen. Microbiol.* 60: 181.

Heath IB and Greenwood AD. 1970. The structure and formation of Iomasomes. *J. Gen. Microbiol.* 62: 129.

Hebraud M and Fevre M. 1987. Protoplast production and regeneration from mycorrhizal fungi and their use for isolation of mutants. *Can. J. Microbiol.* 34: 157-161.

Heredia CJ, Sols A and de la Fuenteg. 1968. *Eur. J. Biochem*. 5:321-329.

Hideo Toyama, Kohtaro Yamaguchi, Atsuhiko Shinmyo and Hirosuke Okada, 1984. Protoplast fusion of *Trichoderma reesei* using immature conidia. *Appl. Environ. Microbiol.* 563-568.

Hinnen A, Hicks JB and Fink GR. 1978. Transformation of yeast. *Proc. Nat. Acad. Sci.* USA, 75: 19-29.

Hocart MJ, Lucas JA and Peberdy JF. 1987. Production and regeneration of protoplast from *Pseudocercosporella heterotrichoides* (Fron) Deighton. *J. Phytopath.* 119: 193-205.

Hockney RC and Freeman RF. 1980. Construction of a polysaccharide degrading brewing yeast by protoplast fusion. In: Advances in Protoplast Research Ferenczy L and Farkas GL (Eds), Akademiai Kiado, Budapest, Pergamon Press, Oxford, pp. 139-143.

Hoekstra D and Kok JW. 1989. Entry mechanisms of enveloped viruses. Implications for fusion of intracellular membranes. *Biosci. Rep.* 9: 273-305.

Holter H and Ottolenghi P. 1960. *C.r. Trav. Lab.Carlsberg*. 31: 409-422.

Homolka L, Vyskocil P and Pilat P. 1988. Use of protoplast in the improvement of filamentous fungi. Mutagenization of protoplasts of *Oudemansiella mucida. App. Microb. Biotech.* 28: 166-169.

Hope MJ and Cullis PR. 1981. The role of nonbilayer lipid structures in the fusion of human erythrocytes induced by lipid fusogens. *Biochem. Biophys.* Acta. 640: 82-90.

Hopwood D. 1981. Genetic studies with bacterial protoplasts. *Ann. Rev. Microbiol.* 35: 237-272.

Hopwood DA, Kieser T, Wright HM and Bibb MJ. 1983. Plasmids, recombination and chromosome mapping in *Streptomyces lividans 66, J. Gen. Microbiol.* 129: 2257-2269.

Hopwood DA, Bibb MJ, Chater KF, Kieser T, Bruton CJ, Kieser HM, Lydiate DJ, Smith CP, Ward JM and Schrempf H. 1985. Genetic manipulation of *Streptomyces*: A laboratory manual, Norwich, John Innes Foundation UK.

Housset P, Nagy M and Schwenke J. 1975. Protoplasts of *Schizosaccharomyces prombe*: an improved method for their preparation and the study of their guanine uptake. *J. Gen. Microbiol.* 90: 260-264.

Hrmova M and Selitrennikoff CP. 1987. Protoplasts of *Neurospora crassa* by an inducible enzyme system of *Arthrobacter*. GJM. I. *Curr. Microbiol.* 16: 33-38.

Huang HC and Hoes JA. 1976. Penetration and infection of *Sclerotinia sclerotiorum* by *Contothyrium minitans. Can. J. Bot.* 54: 406-410.

Huang D, Staples RC, Bushnell WR and Maclean DJ. 1990. Preparation and regeneration of protoplasts from axenic mycelia derived from the wheat stem rust fungus. *Phytopath.* 80: 81-84.

Hutchinson HT and Hartwell LH. 1967. Macromolecule synthesis in yeast spheroplasts. *J. Bact.* 94: 1697-1705.

Hutter R and Eckhardt T. 1988. Genetic manipulation In: Actinomycetes in Biotechnology, Goodfellow M, Williams ST and Mordanski M (Eds), Academic Press, London, pp. 90-184.

Iijima Y and Yangi SO. 1986. A method for high yield preparation and high frequency regeneration of basidiomycete, Pleurotis ortreatus (Hirabke), protoplasts using sulphite pulp waste components. *Agricultural and Biological Chem.* 50: 1855-1861.

Ingolia TD and Wood JS. 1986. Genetic manipulation of *Saccharomyces cerevisiae*. In: Manual of Industrial Microbiology and Biotechnology, (Demain AL, Solomon NA (Eds), American Society Washington, DC for Microbiologists, pp. 204-213.

Issac S. 1978. Biochemical properties of protoplasts from *Aspergillus nidulans*. Ph.D. thesis. Univ. of Nottingham, England. 283 pp.

Issac S and Gokhale DV. 1982. Autolysis: A tool for production from *Aspergillus nidulans. Trans. Brit. Mycol. Soc.* 78: 389-394.

Jackson CW and Heale JB. 1987. Parasexual crosses by hyphal anastomosis and protoplast fusion in the entomopathogen *Verticillium lecanii. J. Gen. Microbiol.* 133: 3537-3548.

Jacobsen T, Jense BO, Olsen J and Allermann K. 1985. Preparation of protoplasts from mycelium and anthroconidia of *Geotrichum candidum. Can. J. Microbiol.* 31: 93-96.

Janderova B, Cvrckova R and Bendova O. 1990. Construction of the dextrin degrading of brewing yest by protoplast fusion. *J. Basic Microbiol.* 30: 499-505.

Javadekar VS, Sivaraman H and Godhale DV. 1995. Industrial yeast strain improvement. Construction of a highly flocculent yeast with a killer character by protoplast fusion. *J. Ind. Microb.* 15: 94-102.

Kakar SN, Partridge RM and Magee PT. 1983. A genetic analysis of *Candida albicans*: isolation of a wide variety of auxotrophs and demonstration of linkage and complementation. Genetics 104: 241-255.

Kaneko H and Sakachuchi K. 1979. Fusion of protoplasts and genetic recombination of *Brevibacterium flavum. Agric. Biol. Chem.* 43: 1007-1013.

Karpagam S. 1994. Variations in protoplast regenerated isolates of *Trichoderma harzianum*. M. Phil. dissertation, University of Madras, India.

Kalpana R. 1995. Evaluation of protoplast regenerated cultures of *Colletotrichum capsici* (Syd.) Butler and Bisby for growth, virulence and control measures. M. Phil. dissertation, University of Madras, India, Pages 27-29.

Kao KN and Michayluk MR. 1974. A method for high-frequency intergeneric fusion of plant protoplasts. *Planta* 115: 355-367.

Kawakami N and Kawakami H. 1971. *J. Ferment. Tech.* 49: 479-487.

Kawamoto H and Aizawa K. 1986. Fusion conditions for protoplasts of an entomogenous fungus *Beauveria bassiania. Appl. and Entomol. zoology* 21: 624-626.

Kelkar HS, Shanker V and Deshpande MV. 1990. Rapid isolation and regeneration of *Sclerotium rolfsii* protoplasts and their potential application for starch hydrolysis. *Enzyme Microb. Technol.* 12: 510-514.

Keller WA and Melchers G. 1973. The effect of high pH and calcium on *Tobacco* leaf protoplast fusion. *Z. Naturforsch.* 28: 737-741.

Keller U. 1983. Highly efficient mutagenesis of *Claviceps purpurea* by using protoplasts. *Appl. Environ. Microbiol.* 46: 580-584.

Kevei F and Peberdy JF. 1979. Induced segregation in interspecific hybrids of *Aspergillus nidulans* and *A. rugulosus* obtained by protoplast fusion. *Mol. Gen. Genetics.* 170: 213-218.

Kevei F and Peberdy JF. 1984. Interspecies hybridization following protoplast fusion in *Aspergillus*. In: Fungal protoplasts: Their use in physiology, biochemistry and genetics, Peberdy JF and Ferenczy L (Eds), Marcel Dekker, New York.

Kevi F and Peberdy JF. 1977. Interspecific hybridization between *Aspergillus nidulans* and *Aspergillus rugulosus* by fusion of somatic protoplasts. *J. Gen. Microbiol.* 102: 255-262.

Kevi F and Peberdy JF. 1985. Interspecies hybridization after protoplast fusion in *Aspergillus*. In: Fungal protoplasts: Applications in Biochemistry and Genetics, Peberdy JF and Ferenczy L(Eds), Marcel Dekker, New York pp. 241-257. Isolation, culture and reversion of protoplast of valvarietta diplacia.

Khanna, PK, Kaur J, Garcha HS, Khanna V, Dhaliwal RPS and Mittar D. 1991. Isolation, culture and oression of Protoplast of *Valvariella diplacia*. Proceedings of the National Symposium on mushrooms, pp. 148-153.

Kiguchi T and Yanagi SO. 1985. Intraspecific heterokaryon and fruit body formation in *Coprinus macrorhizus* by protoplast fusion of auxotrophic mutants. *Appl. Microbiol. Biotechnol.* 22: 121-127.

Kim KS, Ryu DDY and Lee SY. 1983. Application of protoplast fusion technique to genetic recombination of *Micromonospora rosaria. Enzyme Microb. Technol.* 5: 273-280.

Kinsky SC. 1962. *J. Bact.* 83: 351-358.

Kinsky SC. 1963. *Archs. Biochem. Biophy.* 102: 180-188.

Kirimura K, Nakajima J, Lee SP, Kawabe S and Usami S. 1988. Citric acid production by the diploid strains of *Aspergillus niger* obtained by protoplast fusion. *Appl. Microbiol. Biotechnol.* 27: 504-506.

Kirimura K, Imura M, Kato Y, Lee SP and Usami S. 1988. Intergeneric protoplast fusion between *Aspergillus niger* and *Trichoderma viride*. *Agric. Biol. Chem.* 52(5): 1327-1329.

Klinner U, Bottcher F and Samsonova I.A. Hybridization of *Pichia guilliermondii* by protoplast fusion. (This volume)

Knutton S. 1979. Studies on membrane fusion. III. Fusion of erythrocytes by Sendai virus. *J. Cell Sci.* 28: 189-210.

Kobori H, Takata, Y and Osumi M. 1991. Interspecific protoplast fusion between *Candida tropicalis* and *Candida boidinii*: characterization of the fusants. *J. Ferment. Bioeng.* 72: 439-444.

Kohno M, Tanaka H, Ishizaki H and Kunoh H. 1983. *Ann. Phytopath. Soc.* Japan.

Komei Fukui, Yasunobu Sagara, Nagayuki Yoshida and Toshiro Matsuoka. 1969. Analytical studies on regeneration of protoplasts of *Geotrichum candidum* by quantitative thin-layer-agar plating. *J. Bact.* 98: 256-263.

Kopecka M, Phaff HJ and fleet GH. 1974. Demonstration of a fibrillar component in the cell wall of the yeast *Saccharomyces cerevisiae* and its chemical nature. *J. Cell Biol.* 67: 66-76.

Kovac L, Bednarova H and Gresak M. 1968. *Title Biochem. Biophys. Acta* Title 153: 32-42.

Kreger DR and Kopecka M. 1975. On the nture and formation of the fibrillar nets produced by protoplasts of *Saccharomyces cerevisiae* in liquid media: An electron microscopic, X-ray diffraction and chemical study. *J. Gen. Microbiol.* 92: 207-220.

Kropp BR and Fortin JA. 1986. Formation and regeneration of protoplasts from the ectomycorrhizal basidiomycete *Laccaria bicolor*. *Can. J. Bot.* 64: 1224-1226.

Kuwabara H, Yumi Magar, Yutaka Kashiwagi, Gentaro Okada and Takashi Sesaki. 1989. Characterization of enzyme productivity of protoplast regenerants from the cellulase producing fungus *Robillarda* Y 20. *Enzy. Microb. Technol.* 11: 696-699.

Kumari DSSVR and Lalithakumari D. 1987. Morphological, physiological an biochemical comparisons between parent (sensitive) and mutant (ediphenphos resistant) strains of *Pyricularia oryzae*. Abst. *Pesticides in Tropical Agriculture*. P.32.

Kumari J and Panda T. 1993. Improved technique on reversion of mycelial protoplast of *Trichoderma reesei* into mycelia. *Bioprocess Eng.* 8: 287-294.

Kuo SC and Lampen JO. 1971. Osmotic regulation of invertase formation and secretion by protoplasts of *Saccharomyces. J. Bact* 106: 183-191.

Laborda F, Garcia-Acha I and Villanueva JR. 1974. *Title Trnas. Br. Mycol. Soc.* 62: 509-518.

Lakshmi BR and Chandra TS. 1993. Rapid release of protoplasts from *Eremothecium Ashbyii* in comparison with *Trichoderma reesei* and *Penicillium chrysogenum* using novozym and funcelase. *Enzyme Microb. Technol.* 15: 699-702.

Lalithakumari D. 1994. Resistance to sterol biosynthesis inhibitors (Baycor and Sisthane) against *Venturia inaequalis*, ICAR Report, New Delhi,

Lalithakumari D, Mrinalini C, Chandra AB and Annamalai P. 1996. Strain improvement by protoplast fusion for enhancement of biocontrol potential integrated with fungicide tolerance in Trichoderma spp. J. of *Plant diseases and Protection* 103: 206-212.

Lalithakumari D. 1996. Protoplasts—A biotechnological tool for plant pathological studies. *Indian Phytopath.* 49: 199-212.

Lalithakumari D and Annamalai P. 1988. Ediphenphos resistance in *P. oryzae* and *D. oryzae in in vitro* techniques for detection and biochemical studies. Chapter 17. In *Managing resistance to agrochemicals* Eds. Maurice B. Green, Homer M. Le Baron, William K. Moberg, Publishers American Chemical society.

Lalithakumari D and Annie Juliet G. 1998. Unpublished data.

Lalithakumari D. 1999. Unpublished data.

Lampen JO Arnow PM, Borowska Z and Laskin AI. 1962. *J. Bact.* 84: L1152-1160.

Landman OE and Halle S. 1963. Enzymically and physically induced inheritance changes in *Bacillus subtilis. J. Mol. Biol.* 7: 721-738.

Lau WC, Dhillon EKS and Chang ST. 1985. Isolation and reversion of protoplasts of pleurotus sajor-caju. *Hircen Journal* 1: 426-428.

Laurila HO, Nevalainen H and Makinen V. 1985. Production of protoplasts from *Curvularia inaequalis* and *Trichoderma reesei. Appl. Microbiol. Biotechnol.* 21: 210-212.

Lazar GB. 1983. Recent developments in plant protoplast fusion and selection technology. *Experiential supply.* 46: 61-68.

Leach J and Yoder OC. 1982 Heterokaryosis in *Cochliobolus heterostrophus. Exp. Mycol.* 6: 364-374.

Lee YH, Park YH, Yoo YB and Min KH. 1986. Studies on protoplast isolation of *Pleurotus cornucopiae. Kor. J. Mycol.* 14: 141-148.

Leikin SL, Kozlov MM, Chernomodik LV, Markin VS and Chizmadzkev YA. 1987. Membrane fusion: overcoming of the hydration barrier and local restructuring. *J. Theor. Biol.* 129: 411-425.

Lesile JF. 1982. Some genetic techniques from *Gibberella zeae. Phytopath.* 73: 1005-1008.

Levitin MM, Konavalova GS and Lebskil VK. 1984. Production, regeneration and fusion of protoplasts in*Drechslera teres Mikologie Fitopathologie* 18: 276-280.

Lilliehoj EB and Ottolenghi P. 1967. In Symposium uber Hefe-Protoplasten, Jena. Ed. R. Muller, Academic-Verlag, Berlin.

Longley RP, Rose AH and Kniths BA. 1968. *Biochem. J.* 108: 401-412.

Lopez-Belmonte F, Garcia Acha I and Villanueva JR. 1966. Observations on the protoplasts of *Fusarium culmorum* and on their fusion. *J. Gen. Microbiol.* 45: 127-134.

Lowe DA and Elander RP. 1983. Contributions of mycology to the antibiotic industry. *Mycologia.* 75: 361-373.

Lucas JA, Greer G, Oudemans PV and Coffey MD. 1990. Fungicide sensitivity in somatic hybrids of *Phytophthora capsici* obtained by protoplast fusion. *Physiol. Mol. Plant Pathol.* 36: 175-187.

Lucy JA, Ahkong QF, Cramp FC, Fisher, D and Jowell JI. 1971. Cell fusion without viruses *Biochem. J.* 124: 469-478.

Lucy JA and Ahkong QF. 1986. An osmotic model for the fusion of biological membranes. *FEBS Lett.* 199: 1-11.

Lurquin PF. 1979. Entrapment of plasmid DNA by liposomes and their interaction with plant protoplasts. Nucleic acids Res. 6: 3773-3784.

Lynch PT, Collin H and Issac S. 1985. Isolation and regeneration of protoplasts from *Fusarium triticum* and *F. Oxysporum. Trans. Br. Mycol.* Soc. 85: 135-140.

MacDonald KD, Hutchinson JM and Gillett WR. 1963. Isolation of auxotrophs of *Penicillium chrysogenum* and their penicillin yields. *J. Gen. Microbiol.* 33: 365-374.

Magae Y, Kakimoto Y, Kashiwagi Y and Sesaki. 1985. *J. Appl. Environ. Microbiol* 49: 441-442.

Maggio B, Ahkong QF and Lucy JA. 1976. Poly(ethylene glycol), surface potential and cell fusion. *Biochem.* J. 158: 647-650.

Makins JF and Holt G. 1981. Liposome mediated transformation of streptomycetes by chromosomal DNA. *Nature* 293: 671-673.

Mamol EA, Mortin FN and Kistler HC. 1989. Isolation and fusion of protoplasts from *F. oxysporum* f. sp. *conglutinans* f. sp. *raphani. Phytopath.* 79: 1204.

Manocha MS. 1968. Electron microscopy of the conidial protoplasts of *Neurospora crassa. Can. J. Bot.* 46: 1561-1567.

Maraz A, Kiss M and Ferenczy L. 1978. Protoplast fusion in *Saccharomyces cerevisiae* strains of identical and opposite matting types. *FEMS Microbiol. Lett.* 3: 319.

Maraz A and Subik J. 1981. Transmission and recombination of mitochondrial genes in *Saccharomyces cerevisiae* after protoplast fusion. *Mol. Gen. Genet.* 181: 131-133.

Matile Ph. 1969. In Symposium on membranes-structure and function, 6th FEBS meeting, Madrid. Eds. JR Villanueva and F Ponz, Academic Presses London.

Martin HH. 1983. Protoplasts and spheroplasts of gram-negative bacteria with special emphasis on *Proteus mirabilis. Experientia Supp.* 46: 213-226.

Marini F, Arnow PM and Lampen JD. 1961. *J. Gen. Microbiol.* 24: 51-62.

Marriott AC, Archer SA and Buck KW. 1984. Mitochondrial DNA in *Fusarium oxysporum* is a 46.5 kilobase pair circular molecule. *J. Gen. Microbiol.* 130: 3001-3008.

Martinkova L, Musilkova M, Ujcova E, Machek F and Seichert L. 1990. Protoplast fusion in *Aspergillus niger* strains accumulating citric acid. *Folia Microbiol* 35: 143-148.

Matsushima P and Baltz RH. 1986. Protoplast fusion, In: Manual of Industrial Microbiology and Biotechnology, Demain AL and Solomon NA (Eds), American Society for microbiology and Washington DC. pp. 170-183.

Matsushima P, McHenney MA and Baltz RH. 1987. Efficeint transformation of Amycolatopsis orientalis (*Nocardia orientalis*) protoplasts by *Streptomyces plasmids. J. Bact.* 169: 2298-2300.

Mellon FM. 1985. Protoplast fusion and hybridization in *Penicillium.* In: Molecular genetics of filamentous fungi, Timberlake WE (Ed) Alan R Liss, New York pp. 69-82.

Montane MH, Dupille E, Alibert G and Teissie J. 1990. Induction of a long-lived fusogenic state in viable plant protoplasts permeabilized by electric fields. *Biochem. Biophys. Acta* 1024: 203-207.

Moore RT and McAlear JH. 1961. Phytopath Z. 42: 297.

Moore PM and Peberdy JF. 1976a. Release and regeneration of protoplasts from the conidia of *Aspergillus flavus. Trans. Br. Mycol. Soc.* 66: 421-425.

Moore PM and Peberdy JF. 1976b. Aparticulate chitin synthase from *Aspergillus flavus* (Link) the properties, location and levels of activity in mycelium and regenerating protoplast preparations. *Can. J. Microbiol.* 22: 915-921.

Morehart AL, Blits KC and Melchior GL. 1985. Pathogenic potential or *Verticillium alboatrum* protoplasts. *Mycologia* 77: 784-790.

Morgan AJ, Heritage H and Whittaker PA. 1977. Protoplast fusion between petite and auxotrophic mutants of the petitenegative yeast, *Kluyveromyces lactis. Microbiol. Lett.* 4: 103.

Morgan AJ. 1983. Yeast strain improvement by protoplast fusion and transformation. *Experientia Suppl.* 46: 155-166.

Mrinalini C and Lalithakumari, 1993. Enhancement of antagonistic and fungicide tolerance potential of *Trichoderma* spp. by protoplast fusion. Abstract at the XV International Botanical Congress, Yokohama, Japan.

Mrinalini C and Lalithakumari D. 1998. Integration of enhanced biocontrol efficacy and fungicide tolerance in *Trichoderma* spp. by electrofusion. *J. Plant.Dis. Prot.* 105: 34-40.

Mrinalini C. 1997. Strain improvement for enhancement and integration of biocontrol potential and fungicide tolerance in *Trichoderma* spp. Ph.D. Thesis, University of Madras, India 81-83.

Mukherjee M and Sen Gupta S. 1988. Isolation and regeneration of protoplasts from *Termitomyces clypeatus. Can. J. Microbiol.* 34: 1330-1332.

Munoz-Rivas A, Specht CA, Drummond BJ, Froeliger E, Noronty CP and Ullrich RC. 1988. Transformation of the basidiomycete *Schizophyllum commune. Mol. Gen. Genet* 205: 103-106.

Musilkova M and Z Fencl. 1968. *Fol. Microbiol.* 13: 235-239.

Muthukumar G, Kulkarni R and Nickerson KW. 1985. Calmodulin levels in yeast and mycelial phases of *Ceratoscystis ulmi. J. Bac.* 162: 47-49.

Nagata T. 1978. A novel cell-fusion method of protoplasts by polyvinyl alcohol. *Naturwissenschaften* 65: 263.

Nagata T, Eibl H and Melchers G. 1979. Fusion of plant protoplasts induced by positively charged synthetic phoshpolipids. Z. Naturforsch. 34c: 460-462.

Nagata T. 1984. Interaction of liposomes and protoplasts as a model system of protoplast fusion. In: Cell Fusion: Gene transfer and transformation Beers RF (Jr.) and Bassett EG (Eds), Raven Press, New York, pp. 217-226.

Necas O. 1971. Cell wall synthesis in yeast protoplasts. *Bacteriological Reviews* 35: 149-170.

Nehls R. 1978. The use of metabolic inhibitors for the selection of fusion products of higher plant protoplasts. *Mol. Gen. Genet.* 166: 117-118.

Neumann E and Rosenheck K. 1972. Permeability changes induced by electric impulses in vesicular membranes. *J. Membr. Biol.* 10: 279-290.

Nurminen T, Oura E and Suomalainen H. 1970. *Biochem. J.* 116: 61-69.

Ochiai-Yanagi S, Monma M, Kawasumi T, Hino A, Kito M and Takabe I. 1985. Conditions for isolation of and colony formation by mycelial protoplasts of *Coprinus macrorrhizus. Agric. Biol. Chem.* 49: 171-179.

Ogawa K, Brown JA and Wood TM. 1987. *Enzy. Microb. Technol.* 9: 229.

Ogawa K, Ohara H, Koide T and Toyama N. 1989. Intraspecific hybridization of *Trichoderma reesei* by protoplast fusion. *J. Ferment. Bioeng*. 67: 207-209.

Okada Y, Suzuki T and Hosaka Y. 1957. Interaction between influenza virus and Ehrilich's tumor cells. III. Fusion phenomenon of Enrlich's tumor cells by the action of HVJ strain. *Med. J. Osaka Unvi.* 7: 709-717.

Okanishi M, Suzuki K and Umezawa H. 1974. Formation and regeneration of streptomycete protoplasts: cultural conditions and morphological study. *J. Gen. Microbiol.* 80: 389-400.

O'Malley KA and Davidson RL. 1977. A new in suspension fusion techniques with polyethylene glycol. *Somat. Cell Genet*. 3: 441-448.

Ossanna N and Mischike S. 1990. Genetic transformation of the biocontrol fungus *Gliocladium virens* to benomyl resistance. *Appl. Environ. Microbiol.* 56: 3052-3056.

Palleroni NJ. 1983. Genetic recombination in *Actinoplanes brasiliensis* by protoplast fusion. *Appl. Environ. Microbiol*. 45: 1865-1869.

Partridge BM and Drew JA. 1974, *Sabouraudia.* 12: 166-178.

Peberdy JF. 1976. Isolation and properties of protoplasts from filamentous fungi. In: Microbial and Plant protoplasts Peberdy JF, Rose AH, Rogers HJ and Cocking EC (Eds), Academic Press, London, pp. 39-50.

Peberdy JF. 1979. Fungal protoplasts: Isolation, reversion and fusion. *Ann. Rev. Microbiol.* 33: 21-29.

Peberdy JF. 1980. Protoplast fusion A New Approach to Interspecies Genetic Manipulation and Breding in Fungi. In Advances in Protoplast Research. Eds. Ferenzy I and Farkas GL. Proc. Of the 5th International Protoplast Symposium, Pergamon, Oxford.

Peberdy JF. 1983. Genetic recombination in fungi following protoplast fusion and transformation. In: Fungal differentiation, Smith JE (Ed.), Marcel Dekker, New York, pp. 559-581.

Peberdy JF and Buckley CE. 1973. Adsorption of fluorescent brighteners by regenerating protoplasts of *Aspergillus nidulans. J. Gen. Microbiol.* 74: 281-288.

Peberdy JF and Gibson RK. 1971. Regeneration of *Aspergillus nidulans* protoplasts. *J. Gen. Microbiol.* 69: 325-330.

Peberdy JF and Isaac S. 1976. An improved procedure for protoplast isolation in *Aspergillus nidulans. Microbios Lett.* 3: 7-9.

Peberdy JF Buckley CE, Daltrey DC and Moore PM. 1976. Factors affecting protoplast release in some filamentous fungi. *Trans. Brit. Mycol. Soc.* 67: 23-26.

Peberdy JF, Eyssen H and Anne J. 1977. Interspecific hybridization between *Penicillium chrysogenum* and *Penicillium cyaneo-fulvum* following protoplast fusion. *Molec. Gen. Genet.* 157: 281.

Peberdy JF and Ferenczy L. 1985. Fungal protoplasts: Applications in Biochemistry and Genetics. Mycology series 6. Marcel Dekker Inc., New York, pp 354.

Peberdy JF. 1989. Fungi without coats-protoplasts as tools for mycological research. *Mycol. Res.* 93: 1-20.

Pfeifer TA and Khachatourians GG. 1987. The formation of protoplasts from *Beauveria bassiana. Appl. Microbiol. Biotechnol.* 26: 248-253.

Picataggio SK, Schamhart DHJ, Mountenecourt BS and Eveleigh DE. 1983. Spheroplast formation and regeneration in *Trichoderma reesei. European J. Appl. Microbiol. Biotechnol.* 17: 121-128.

Pontecorvo G and Roper JA. 1952. Genetic analysis without sexual reproduction by means of polyploidy iin *Aspergillus nidulans. J. Gen. Microbiol.* 6: vii.

Pontecorvo G. 1956. The parasexual cycle in fungi. *Ann. Rev. Microbiol.* 10: 393-400.

Potrykus F, Harms CT, Hinnen, Huter R, King PJ and Shilleto RD. 1983. Protoplasts. *Experientia Suppl.* 46: 1-269.

Power JB, Cummins SE and Cocking EC. 1970. Fusion of isolated plant protoplasts. *Nature* 225: 1016-1018.

Pratsch L, Herrmann A, Schwede I and Meyer HW. 1989. The influence of polyethylene glycol on the molecular dynamics within the glycocalyx. *Biochem. Biophys. Acta.* 980: 146-154.

Preece TF. 1971. Fluorescent techniques in mycology. In: Methods in Microbiology, Vol. 4: Ed. Booth Ed. C Academic press, London pp. 509-516.

Radford A, Susan Pope, Sazci A, Madeline J Fraser and Parish JH. 1981. Liposome-mediated genetic transformation of *Neurospora crassa. Mol. Gen. Genet.* 184: 567-569.

Rand RP. 1981. Interacting phospholipid bilayers: measured forces and inducted structural changes. *Ann. Rev. Biophys. Bioengg.* 10: 227-314.

Raymond P, Veau P and Fevre M. 1986. Production by protoplast fusion of new strains of *Penicillium* caseicolum for use in the dairy industry. *Enzyme Microb. Technol.* 8: 45-48.

Revathi R. 1993. Molecular and genetic perspectives of fungicide (Bitertanol and Fenapanil) resistnace in *Venturia inaequalis* (Cooke) Wint. Ph.D. Thesis.

Revathi R and Lalithakumari D. 1992. *Venturia inaequalis*: A novel method for protoplast isolation and regeneration. *Z. Pflanzenkrankh. Pflanzensch.* 100: 211-219.

Robinson JM, Roos DS, Davidson RL and Karnovsky MJ. 1979. Membrane alterations and other morphological features associated with polyethylene glycol-induced cell fusion. *J. Cell Sci.* 40: 63-75.

Rodicio ML, Manzanal MB and Hardisson C. 1978. Protoplast like structures' formation from two species of enterobacteriaceae by fostomycin treatment. *Arch. Microbiol,* 118: 219-221.

Rodriguez Aguirre MJ, Garcia Acha I and Villanueva JR. 1964. Formation of protoplasts of *Fusarium culmorum* by strepzyme. *Antonie van Leeuwenhoek* 30: 33.

Rost K and Venner H. 1965. Wachstum Von Hefeprotoplasten und ihre Reversion Zu Intakten Helfezellen. *Archiv. Microbiol.* 51: 130.

Ruiz-Herrera J and Bartnicki-Garcia S. 1976. Proteolytic activation and inactivation of chitin synthetase from *Mucor rouxii. J. Gen. Microbiol.* 97: 241-249.

Russel I and Stewart GG. 1979. Spheroplast fusion of brewer's yeast strain. *J. Inst. Brew* London 85: 95-98.

Russel I, Jones RM, Weston BJ and Stewart GG. 1983. Liposome-mediated DNA transfer in brewing and related yeast strains. *J. Inst. Brew.* 89: 136.

Sagara Y. 1969. *Tokushima J. Exp.* Med. 16: 57-69.

Sakaguchi K, Ochi K, Gunge N and Uchida K. 1980. Protoplast fusion. In: Molecular Breeding and Genetics of Applied Microorganisms, Sakaguchi K and Okanishi M (Eds), Academic press, New York, pp. 85-105.

Salek A, Schnittler R and Zammerman U. 1992. Stably inherited killer activity in industrial yeast strains obtained by electrotransformation. *FEMS Microbiol. Lett.* 96: 103-110.

Sandhu DK, Wadhwa Vipul and Bagga PS. 1989. Use of lytic enzyme for protoplast production in *Trichoderma reesei. Enzyme Microbiol.* Technol. 11: 21-25.

Sandri-Goldin RM, Goldin AL, Levine M and Glorioso J. 1983. High-efficiency transfer of DNA into eukaryotic cells by protoplast fusion. *Methods Enzymol.* 101: 402-411.

Santiago CM. 1983. Volvariella Volvacea protoplasts: Isolation, reversion and fusion. In: UNESCO Regional workshop on protoplast fusion in Microorganisms, Seoul (Korea) Aug. 16-22, 1983.

Santos TMC and De Melo IS. 1991. Preparation and regeneration of protoplasts of *Talaromyces flavus. Rev. Bras. Genet.* 14: 335-340.

Sarachek A, Rhoads DD and Schwarzhoff RH. 1981. Hybridization of *Candida albicans* through fusion of protoplasts. *Arch. Microbiol.* 129: 1-8.

Savchenko GV and Kapultsevich Yug. Hybridization of yeast from genus *Hansenula* by protoplast fusion. (This volume)

Schlenk F and Dainko JL. 1966. *J. Bact.* 89: 428-436.

Schmid EN. 1984. Fosfomycin-induced protoplasts and L-forms of *Staphylococcus aureus*. *Chemotherapy* 30: 35-39.

Schupp T and Divers M. 1986. Protoplast preparation and regeneration in *Nocardia mediterranei FEMS Microbiol. Lett.* 36: 159-162.

Sentandroll *et al.* 1966.

Shahin MM. 1972. *J. Bact.* 110: 769-771.

Shimizu S. 1987. Protoplast fusion of insect pathogenic fungi. In: Biotechnology in invertebrate pathology and cell culture, Maramorosch K. (Ed.), Academic press, San Diego, pp 410-414.

Shivla GF, Schlenk JL and Dainko. 1961. *J. Bacteriol.* 82: 808-814.

Shockman GD and Lampen J O. 1962. *J. Bacteriol.* 84: 508-512.

Sutton DD and lampen JO. 1962. *Biochim. Biophys. Acta.* 56: 303-312.

Siegel DP. 1986. Inverted micellar intermediates and the transitions between lamellar, cubic and inverted hexagonal lipid phases. *Biophys.* J. 49: 1171-1183.

Sietsma JH and De Boer WR. 1973. Formation and regeneration of protoplasts of *Pythium* PRL 2142. *J. Gen. Microbiol.* 74: 211-217.

Sietsma JH, Eveleigh DE and Haskins RH. 1968. Title*Antonie van Leeuwenhoek* 34: 331-340.

Silveria WD and Azevedo JL. 1987. Protoplast fusion and genetic recombination in *Metarhizium anisopliae. Enzyme Microb. Technol.* 9: 149-151.

Sipiczki M and Ferenczy L. 1977. Protoplast fusion of *Schizosaccharomyces pombe* auxotrophic mutants of identical mating type. *Mol. Gen. Genet.* 151: 77-81.

Sivaprakash KS and Bateen C. 1981. Segregation and transmission of mitochondrial makers in fusion products of the asporogenus *Torulopsis giabrata. Curr. Genet.* 4: 73-80.

Sivan A, Harman Ge and Stasz TE. 1990. Transfer of isolated nuclei into protoplasts of *Trichoderma harzianum. Appl. Environ. Microbiol.* 56: 2404-2409.

Smith AW, Collins K, Ramsden M, Fox HM and Peberdy JF. 1991. Chromosome rearrangements in improved cephalosporin C producing strains of *Acremonium chrysogenum. Curr. Genet.* 19: 235-237.

Soliday CL, Dickman MB and Kolattukudy PE. 1989. Structure of the cutinase gene and detection of promoter activity in the 5 flanking region by fungal transformation. *J. Bact.* 171: 1942-1951.

Sommer A and Lewsi MJ. 1971. *J. Gen. Microbiol.* 69: 327-335.

Sonnenberg ASM, Sietsma JH and Wessels JGH. 1982. Biosynthesis of alkali-insoluble cell wall glucan in *Schizophyllum commune protoplasts.* 128: 2667-2674.

Spata L and Weber H. 1980. A study on protoplast fusion and parasexual hybridization of alcne utilizing yeasts. (This volume).

Spencer JFT and Spencer DM. 1983. Genetic improvement of industrial yeasts. *Ann. Rev. Microbiol.* 37: 1221-1242.

Spencer JFT and Spencer DM. 1980. The use of mitochondrial mutants in the isolation of hybrids involving industrial yeast strains. *Mol. Gen. Genet.* 177: 355-358.

Srikantha T and Rao GR. 1984. A method for the isolation of protoplasts from Dermatophytes. *J. Gen. Microbiol.* 130: 1503-1506.

Stahl U. 1978. *Mol. Gen. "Genet.* 160: 111-113.

Stanway CA and Back KW. 1984. Infection of protoplasts of the wheat take all fungus *Gaeumannomyces graminis var. tritici* with double stranded RNA viruses. *J. Gen. Virol.* 65: 2061-2065.

Stasz TE Harman GE and Weeder NF. 1988. Protplasts preparation and fusion in two biocontrol strains of *Trichoderma harzianum. Mycologia* 80: 141-150.

Stewart GG. 1981. The genetic manipulation of industrial yeast strains. *Can. J. Microbiol.* 27: 973-990.

Strominger JL. 1968. Enzymatic reactions in bacterial cell wall synthesis sensitive to penicillins and other antibacterial substances. In: Microbial protoplasts, Spheroplasts and L-forms (Guze LB (Ed.), Williams & Wilkins, Baltimore pp. 55-61.

Strunk C. 1969. Licht-und elektronenmikroskopische Untersuchungen an jungen Protoplasten von Polystictus versicolor. *Zeitschrift fur Allgemeine Mikrobiologie*. 9: 49-60.

Sugiyama M, Katoh T. Mochizuki H, Nimi O and Nomi R. 1983. *J. Ferment. Technol.* 61: 347-351.

Svoboda A. 1966. Regeneration of yeast protoplasts in agar gels. *Exptl. Cell. Res.* 44: 640-642.

Svoboda A. 1977. Intra species fusion of yeast protoplasts. *Folia Microbiol.* 22: 441-442.

Svoboda A. 1978. *J. Gen. Microbiol* 109: 169-175.

Svoboda A. 1980. Intergeneric fusion of yeast protoplasts: *Saccharomyces cerevisiae + Schizosaccharomyces pombe*. In: Advances in protoplast Research. Ferenczy L. and Farkas GL (Eds), Akademiai kiado, Budapest, Pergamon Press, Oxford, pp. 119-124.

Svoboda A and Necas O. 1966. *Nature* 210: 845.

Tabata S, Imai T and Termi B. 1965. *J. Ferment. Technol.* 43: 221-227.

Tamai A, Miori K and Kayama T. 1989. Characterization of Protplasts from basidiospores of edible Basidiomycetes *Res. Bul. Coll. Exp. Forests* 46: 425-440.

Tamova G, Betina V and Farkas V. 1993. An efficeint method for the preparation of protoplasts from *Trichoderma viride. Folia Microbiol.* 38: 214-218.

Tanaka H, Ogaswara N and Kamimiya S. 1981. Protoplat of *Pyricularia oryzae* p2: preparatiion and regeneration into hyphal form. *Agr. Biol Chem.* 45: 1541-1552.

Tanaka C, Yasuzuki K and Tsuda M. 1988. Comparison of mutagens *Cochliobolus* heterostrophus mutagenesis. *Ann. Phytopath. Soc.* Japan 54: 503-509.

Tashpulatov ZH, Shulman TS, Baibaev BG and Mirzarakhimowa M. 1991. Protoplast formation and regeneration by cellulolytically active fungus *Trichoderma harzianum* 19. *Microbiologiya* 60: 541-545.

Tebeest DO and Weidermann GJ. 1990. Preparation and regenration of protoplasts of *Colletotrichum gloesporioides* f. sp. *aeschynomene. Mycologia* 82: 249-255.

Teissie J and Rols MP. 1986. Fusion of mammalian cells in culture is obtained by creating the contact between cells after theiir electropermeabilization. *Biochem. Biophys. Res. Commun.* 140: 258-266.

Telesnina GN, Yu O Sazykin, Zvyagilskaya RA, Dmitrieva SV, Krakhmaleva IN, Novikova ND, Yu E Bartoshevich and Navashin SM. 1988. Protoplasts of *Acremonium chrysogenum*: Biochemical and Morphological investigations. *Antibiot Khimioter.* 33: 190-196.

Toister Z and Loyter A. 1971. Ca^{2+} induced fusion of avian erythrocytes. *Biochem. Biophys. Acta* 241: 719-724.

Tomscik J and Guex-Holzer S. 1952. Anderung der Struktur der Bakterienzelle im Verlauf der Lysozym-Einwirkung. *Schweiz. Z. Allgem. Pathol. Bakteriol.* 15: 517-525.

Tonino GJM and Steyn-Parve EP. 1963. *Biochim. Biophys. Acta* 67: 453-459.

Toyama H, Shinmyo A and Okada H. 1983. Protoplast formation from conidia of *Trichoderma reesei* by cell wall lytic enzymes of a strain of *Trichoderma viride*. J. Ferment. Technol. 61: 409-411.

Toyama, H., Yamaguchi K. Shinmya A. and Okada. 1984. Protoplast fusion of *Trichoderma reesei.* Using immature conidia. pppl. Environ. Microbiol. 47: 362-366.

Toyomasu T, Matsumoto T and Mori K. 1986. Interspecific protoplast fusion between *Pleurotus ostreatus and Pleurotus salmneo stamineus. Agri. Biol. Chem.* 50: 223-225.

Trinci APJ. 1978. In: The filamentous fungi. Smith JE and Berry DR (Eds), Edward Arnold, London 3: 28-50.

Turgeon BG Garber RB and Yoder OC. 1985. Transformation of the fungal maize pathogen *Cochliobolus heterostrophus* using the *Aspergillus nidulans* and S gene. *Mol. Gen. Genet.* 201: 450-453.

Tyagi JS, Tyagi AK and Venkitasubramanian TA. 1981. Preparation and properties of spheroplasts from *Aspergillus parasitica* with special reference to the *de novo synthesis* of aflatoxins. *J. Appl. Bacteriol.* 50: 481-491.

Typass MA. 1983. Heterokaryon incompatibility and interspecific hybridization between *Verticillium albo-atrum* and *Verticillium dahlae* following protoplast fusion. *J. Gen. Microbiol.* 129: 3043-3056.

Tyrell D and MacLeod DM. 1972. I. *Invert. Pathol.* 19: 354-360.

Uchida T. 1988. Introduction of macromolecules into mammalian cells by cell fusion. *Exp. Cell Res.* 178: 1-17.

Upchurch RG, Ehrenshaft M, Walker DC and Sanders RS. 1991. Genetic trnsformation system for the fungal soybean pathogen *Cercospora kikuchii. Appl. Environ. Microbiol.* 57: 2935-2939.

Urano N, Nishikawa N and Kamimura N. 1990. Electrofusion of brewer's yeast and *Saccharomyces cerevisiae* (FLO5). In: Proc of the 21st Con. Institute of Brewing (Australia and New Zealand section). Australian Industrial Publ, Adelaide. 153-157.

Urano N, Sahara H and Koshino S. 1993. Conversion of nonflocculent brewer's yeast to flocculent ones by electrofusion. I. Identification and characterization of the fusants by pulsed field electroporation. *J. Biotechnol.* 28: 237-247.

Ushijima S, Nakadai T and Uchida K. 1990. Further evidence on the interspecific protoplast fusion between *Aspergillus oryzae* and *Aspergillus sojae* and subsequent haploidization, with special reference to their production of some hydrolyzing enzymes. *Agric. Biol. Chem.* 54: 2393-2399.

Vallin C and Ferenczy L. 1978. Diploid formation of Candida tropicalis via protoplast fusion. Acta Microbiol. Acad. Sci. Hung. 25: 209.

Van Dam GJW, Slavenburg JH and Koningsverger VV. 1964. *Biochem. J.* 92: 48.

Van Der Valk P and Wessels JGH. 1973. Mitotic synchrony in multinucleate *Schizophyllum* protoplasts. *Protoplasma* 78: 427-432.

Van Der Valk P and Wessels JGH. 1976. Ultrastructure and localization of wall polymers during regeneration and reversion of protoplasts of *Schizophyllum commune. Protoplasma* 90: 65-87.

Van Solingen P and Van der Plaat JB. 1977. *J. Bact.* 130: 946-947.

Veenstra AE, Van Solingen P. Huiningamuurling H, Koekman BP, Groenen MAM Small EB, Kattevilder A, Alvarex E, Barredo JL and Martin JF. 1989. Cloning of penicillin biosynthetic genes. In: Genetics and Molecular Biology of Industrial Microorganisms, Hershberger CL,

Queener SW and Hegeman G (Eds), American Society for Microbiology, Washington, DC, pp. 262-269.

Verkleij AJ. 1984. Lipidic intramembranous particles. *Biochem. Biophys. Acta*. 779: 43-63.

Vijayapalani P. 1995. Biochemial, physiological and molecular aspects of penconazole and carbendazim resistance in mutants and protoplast fusants of *Venturia inaequalis* (Cooke) Wint, Ph.D. Thesis, University of Madras, India 209-213.

Villanueva JR. 1966. Protoplasts of Fungi In The Fungi Eds. GC Ainsworth and AS Sussman. Vol. 2, P.2, New York Academic Press.

Villanueva JR and Acha G. 1971. Production and use of fungal protoplast. In: Methods in Microbiology, Booth G (Ed). Academic Press, New York 4: 666-718.

Villanueva JR. 1966. Protoplasts of fungi. In: the fungi. GC Ainsworth and Sussman AS. (Eds), Vol 2, Academic Press, New York.

Volfova O, Munck V and Dostalek M. 1968. *Experentia* 23: 1005-1006.

Vondrej V. 1987. A killer system in yeasts: applications to genetics and industry. *Microbiol Sci*. 4: 313-316.

Wallin A, Glimelius K and Eriksson T. 1974. The induction and aggregation of *Daucus carota* protoplasts by polyethylene glycol. *Z. Pflazenphysiol*. 74: 64-80.

Wang J, Holden DW and Leong SA. 1988. Gene transfer system for the phytopathogenic fungus Ustilago maydis. *Proc. Natl. Acad. Sci*. USA 85: 865-869.

Watari J, Kudo N and Kamimura M. 1990. Construction of flocculant yeast cells (*Saccharomyces cerevisiae*) by mating or protoplast fusion using a yeast cell containing the flocculation gene FLO5. *Agric. Biol. Chem*. 54: 1677-1681.

Weibull C. 1953. The isolation of protoplasts from *Bacillus megaterium* by controlled treatment with lysozyme. *J. Bact*. 66: 688-695.

Wessels JGH, Kreger DR, Marchant R, Regensburg BA and De Vries OHM. 1972. Chemical and morphological characterization of the hyphal wall surface of the basidiomycete *Schizophyllum commune*. *Biochem. Biophys. Acta*. 273: 346-358.

Whittaker PA and Andrews KR. 1969. *Microbios*. 1: 99-104.

Wiegand R, Weber G, Zimmermann K, Monajembashi S, Wolfrum J and Greulich KO. 1987. Laser-induced fusion of mammalian cells and plant protoplasts. *J. Cell Sci*. 88: 145-149.

Wright EW. 1978. The isolation of heterokaryons and hybrids by a selective system using irreversible biochemical inhibitors. *Exp. Cell Res*. 112: 395-407.

Yabuki M, Kasai Y, Ando A and Fuji T. 1984. Rapid method for converting fungal cells into protoplasts with high regeneration frequency. *Exp. Mycol.* 8: 386-390.

Yamada, O, Magae Y, Kashigana Y, Kakimoto Y and Sasaki T. 1983. Preparation and regeneration of mycelial protoplasts on *Collybia velutipes* and *Pleurotus ostreatus. European J. Appl. Microbiol. Biotechnol.* 17: 298-300.

Yamamoto M and Fukui S. 1977. Fusion of yeast protoplasts, *Agric. Biol. Chem.* 41: 1829.

Yamamura M, Teranishi Y, Tanaka A and Fukui S. 1975. Agric. Biol. Chem. 39: 13-20.

Yanagi SO, Monma M, Kawasumi T, Hino A, Kito M and Takebe T. 1985. Conditions for isolation and colony formation by mycelial protoplasts of *Coprinus macrorhizus. Agri. Biol. Chem.* 49: 171-179.

Yokomori Y, Akiyama H and Shimizu, K. 1989. Breeding of wine yeast through protoplast fusion. *Yeast* 5: S145-S150.

Yoo YB. 1989. Fusion between protoplasts of *Ganoderma applanatum* and oidia of *Lyophyllum ulmarium. Kor. J. Mycol.* 17: 197-201.

Yoo YB, You CH, Shin PG, Park YH and Chang KY. 1987. Transfer of isolated nuclei from *Pleurotus florida* into protoplasts of *Pleurotus ostreatus. Kor. J. Mycol.* 15: 250-253.

Yoshida K. 1979. Interspecific and intraspecific mitochondria-induced cytoplasmic transformation in yeasts. *Plant Cell Physiol.* 20: 851-856.

Yoshida K and Takeuchi IK. 1980. Cytological studies on mitochondrial-induced cytoplasmic transformation in yeasts. *Plant Cell Physiol.* 21: 497-509.

Zimmermann U. 1986. Electrical breakdown electropermeabilization and electrofusion. *Rev. Physiol. Biochem. Pharmacol.* 105: 176-256.

Zimmermann U and Urnovitz BH. 1987. Principles of electrofusion and electropermeabilization. *Methods Enzymol.* 151: 194-221.